Children of Chaos >*

Also by Douglas Rushkoff

Media Virus!: Hidden Agendas in Popular Culture
Cyberia: Life in the Trenches of Hyperspace
The GenX Reader

Children of Chaos >*

[Surviving the End of the World as we Know it]

Douglas Rushkoff

HarperCollins*Publishers*

HarperCollins*Publishers*
77–85 Fulham Palace Road,
Hammersmith, London W6 8JB

Published by HarperCollins*Publishers* 1997
1 3 5 7 9 8 6 4 2

First published in the USA by
HarperCollins*Publishers* 1996
as *Playing the Future*

Copyright © Douglas Rushkoff 1996

Douglas Rushkoff asserts the moral right to
be identified as the author of this work

A catalogue record for this book is
available from the British Library

ISBN 0 00 255626 X

Set in New Aster

Printed and bound in Great Britain by
Caledonian International Book Manufacturing Ltd, Glasgow

For Timothy

CONTENTS

INTRODUCTION

———◆◆———

THE CHILDREN OF CHAOS

> "They keep changing the rules—how we're supposed to behave in each situation. They keep changing it. It's just like the world: everything keeps changing constantly."
>
> U.S. soldier in Haiti police operation, October 1994

Looks like this is the end.

Global warming, racial tension, fundamentalist outbursts, nuclear arsenals, bacterial mutation, Third World rage, urban decay, moral collapse, religious corruption, drug addiction, bureaucratic ineptitude, ecological oversimplification, corporate insensitivity, crashing world markets, paranoid militias, AIDS, resource depletion, hopeless youth, and many, many other indicators of societal health all suggest crisis.

Is it so naïve or even childish to suggest that these may not be signs of doom at all, but only look that way? Couldn't our inability to see our way out of what feels like such a mess be a problem of perception and not design? We might finally be ready, or simply desperate enough, to adopt a paradigm that admits the possibility of something other than decline, decay, and death. If it takes the open-mindedness of youth to develop such a worldview, then so be it.

Those of us intent on securing an adaptive strategy for the coming millennium need look no further than our own children for reassuring answers to the myriad of uncertainties associated with the collapse of the culture we have grown to know and love. Our kids may be younger than us, but they are also newer. They are the latest model of human being, and are equipped with a whole lot of new features. Looking at the world of children is not looking backwards at our own pasts— it's looking ahead. They are our evolutionary future.

Consider any family of immigrants to America. Who learns the language first? Who adopts the aesthetic, cultural, and spiritual values of their new host nation? The children, of course. Educators have concluded that a person's ability to incorporate new language systems—to adapt—is greatest until about the onset of puberty, then it drops off dramatically. As a result, we adults adapt more slowly and less completely than our kids do. Were we immigrants to a new territory, we would be watching our children for cues on how to speak, what to wear, when to laugh, even how to perceive the actions of others.

Well, welcome to the twenty-first century. We are all immigrants to a new territory. Our world is changing so rapidly that we can hardly track the differences, much less cope with them. Whether it's call-waiting, MTV, digital cash, or fuzzy logic, we are bombarded every day with an increasing number of words, devices, ideas, and events we do not understand. On a larger scale, the cultural institutions on which we have grown dependent—organized religion, our leaders and heroes, the medical establishment, corporate employers, even nation states and the family itself—appear to have crumbled under their own weight, and all within the same few decades.

2

Without having migrated an inch, we have, nonetheless, traveled further than any generation in history.

Compare the number of ideas a person is exposed to each day with the number he might have been asked to consider, say, just seventy-five years ago. Inventions like the telephone, television, radio, tickertape, photocopier, fax machine, modem, Internet, cable TV, video teleconferencing, computer bulletin board, and the World Wide Web all function to increase the number of ideas and number of people with whose thoughts we come in contact. With each successive development in communications technology comes a corresponding leap in the number of ideas with which it requires us to cope. As we incorporate each new invention into our daily life, we must accelerate our ability to process new thoughts and ideas.

The degree of change experienced by the past three generations rivals that of a species in mutation. Today's "screenager"—the child born into a culture mediated by the television and computer—is interacting with his world in at least as dramatically altered a fashion from his grandfather as the first sighted creature did from his blind ancestors, or a winged one from his earthbound forebears. Human beings have evolved significantly within a single creature's lifespan, and this intensity of evolutionary change shows no signs of slowing down. What we need to adapt to, more than any particular change, is the fact that we are changing so rapidly. We must learn to accept change as a constant. Novelty is the new status quo.

This naturally contradicts much of what we hold dear. In fact, it calls for us to abandon most of it. Many people still reject the notion that biology evolves, much less that societies or individuals do. But our aim here is to identify a few models for what is happening to our world that point to something other than doom. Evolution may be our best bet.

The evolution of life, so far, appears to be a groping toward complexity. In the face of that pessimistic law called entropy (where everything in the universe slows up and breaks down, cooling off into lifeless, motionless particles) "life" toils to achieve higher states of order through combining smaller parts into larger, networked wholes. Atoms organized themselves into molecules, molecules into more complex, organic chains, which themselves developed into the first single-celled life-forms. These life-forms became the component parts (mitochondria, chloroplasts) of more complex cells, which in turn combined to form organelles that grouped together into organisms. Organisms developed into flocks, herds, and, it can be argued in the case of human beings, into societies and civilizations. The objective, in each case of evolution, was to preserve the organizational state already achieved. The cell preserves the organic chain of molecules, and the herd protects each member.

At each leap of evolution, there has been a corresponding moment where the individual component part was subsumed into something greater than itself. Mitochondria lost a certain aspect of their individuality as unique cells when they became part of the more complex cell of the future. But the greater organism generally has a higher level of awareness and complexity than any of its predecessors.

If we accept that evolution is a process by which matter moves toward higher states of complexity and greater levels of awareness, and does so by developing individuals as far as they can go, and then combining them into new, networked beings, then what is happening to us becomes obvious. We are evolving into a new, colonial life-form.

This process can be scary, especially to members of a culture who value their individuality, personal privacy, and overall stability. While, as we'll see, the development of the "metabeing" need not cost us any of these privileges, we don't have a way yet to understand how we can come together without disappearing ourselves. This is where the experience of a developing child can inform our cultural transition.

4

When a baby is born, it experiences itself as one with its mother. It does not yet understand the concept of its own individuality. The nipple may as well be one of its own organs. There is no "other." Early humans can be likened to the infant. Primitive people experienced themselves as one with the earth, or even one with the tribe. The earth provided food and shelter to the hunter-gatherer, and his primitive, pre-theistic "pagan" religions did not distinguish between himself and his God. Just as, in our own Bible, Adam and Eve walked alongside God in the Garden of Eden, primitive humans were in direct communion with the forces of nature.

Part of a baby's natural development is to realize his separateness. The child learns to speak, and, as any parent will tell you, the most common word in his early vocabulary is "no!" The child celebrates and enforces his individuality. When the child gets to nursery school, the new word is "mine" as the child learns to fight over toys with his peers. Part of what makes all this safe to do is the presence of strong parents who provide boundaries, rules, and the resulting sense of security. The big people serve as much-needed role models for the now self-determining child.

Humankind, too, went through a long journey toward self-awareness. We ceased to see the earth as mother/provider and began exploiting it for its agricultural promise. We developed concepts like property in order to codify what belonged to whom. Specialization, clans, marriage, homes, castes, and classifications of every kind helped us maintain our newfound separateness. Like our child counterparts, we also established "big people" to make us feel safe. Hosts of role models like kings, popes, saints, and lords, both real and mythological, were established to give us the laws and boundaries—the context—we needed to feel safe as individuals. Just like Adam and Eve, we moved out of the Garden and lost our sense of fusion with God and nature. Now God was this guy above us, and we were the children below.

Like the child who fights for his individuality, we developed tools and ideas to preserve our sense of identity. Weapons, dualistic ideologies, reductionist sciences, and self-righteous religions helped us to stave off nature and feel secure about the permanence of our personhood. Unfortunately for us, the children of gods, it is time to grow up.

Long after the "terrible two's," another, more pressing urge than self-promotion strikes the now-pubescent child: sex. The quest for intimacy overwhelms any vestige of the need to assert individuality. As any developmental psychologist will affirm, the battle for independence is actually a necessary step to get to this new, sexual stage. One needs to experience separateness before he can strive for union with someone else. As a person moves into increasing levels of intimacy, he is forced to drop some of his formerly reinforced boundaries. He shares thoughts, space, possessions, and personal fluids with his loved one.

This preadolescent phase is precisely the stage we have found ourselves in as a developing civilization. Just like the child who at first resists the advances of the opposite sex (Gross! Cooties!) we, too, fear the coming of global intimacy, because it means losing a lot of what we've fought so hard to prove in the past. We don't even like to believe that we, as a civilization, are vulnerable to the same sorts of evolutionary laws that dictate the natural selection of individual biological species. We like to think of evolution as something that promotes the dominance of one individual over another. That a bird with a longer beak can get more food from deep inside a tree trunk is a nonthreatening evolutionary presumption. As far as human beings are concerned, most of us will only go so far as to admit that individual people and traits evolve subject to the laws of natural selection: better eyes make for a better hunter, or higher sperm count makes for more plentiful offspring.

To say that culture, too, is subject to the laws of evolution assumes that human society is part of nature. Well, it is. Realizing this fact is half the battle. We invented most of our systems of thought and technological devices to shield us from the harsher realities of nature, but now, ironically, they appear to be forcing us to reckon with them once again. If we can't come to understand the fact that our personal evolution and survival is dependent on species survival, then the battle against our own extinction is already lost.

Like the sexually awakening youngster, we are becoming aware just how intimately linked we all are and must continue to be, whether we feel comfortable about it or not. Our mediating technologies provide the most blatant example of how we are being hardwired together, and it isn't always pretty. Telephones, faxes, computer networks, cable television, satellite links, and cellular services bring a screenager in contact with more people each day than his grandfather could have hoped to meet in his lifetime. This is all pretty nonthreatening when it occurs in the context of education, business teleconferencing, and *Star Trek* conferences, but what about obscene phone calls, junk faxes, information overload, loss of privacy, malicious hackers, computer viruses, corporate espionage, or, worse, virtual communities where new ideas are bandied about without authorities to regulate them? Kids can get hold of pornographic pictures, cancer sufferers can get access to information about untested experimental cures, counterculturalists can network with like-minded malcontents, businesses can learn of our purchasing patterns, and anyone can represent himself online however he chooses.

Meanwhile, the newly formed low-budget bottom-feeders of the mediascape can erode our most established cultural institutions. A pirated phone message tape can shake the foundations of the British monarchy, a camcorder tape can bring down the Los Angeles Police Department, a cable channel can erode our confidence in the U.S. court system, and an impromptu Internet conference can make kids feel

7

better about learning from strangers than from their own teachers.

If, like immature children, we steadfastly maintain our allegiance to the sinking, obsolete institutions of the past, then we will certainly go down with the ship. On the other hand, if we can come to understand this tumultuous period of change as a natural phase in the development of cross-cultural intimacy, then unprecedented pleasure awaits us. We must learn to see the new intrusions into our lives as a teenager experiences flirtatious gestures from an attractive suitor. We are not being invaded—we are just being nudged.

As in any society in crisis, it is the children who first learn to incorporate the worst of threats into the most basic forms of play. When the black plague threatened Europe with anni-hilation, the children sang "Ring Around the Rosie" as they ritualized the appearance of rose-colored sores, stench-camouflaging "posies," and piles of burning bodies in the streets. "Ashes ashes," goes the simple refrain before it con-cludes in straightforward frankness: "We all fall down." How do children today deal with societal crisis of impending sexual intimacy? The answer is to be found at Toys Я Us: they play with slime. The hottest new toys for kids aren't really toys at all, but highly advanced forms of goo. Most adults can remember the craft-oriented goop products of their own childhoods, like Play-Doh and Silly Putty. But these all had apparent purposes. Play-Doh was a reusable modeling clay, and Silly Putty was used to copy and then stretch the impres-sions of comic book pictures. Today's goop equivalents, most notably Gak and Floam, celebrate slime for its own sake, and serve as a great example of the way kids cope first with cul-tural crisis.

The goop revival began in about 1982, when a Nickelodeon show called "You Can't Do That on Television" initiated a practice of dumping a substance called "slime" on its child characters whenever they were at a loss for how to please their adult producer, who had, according to the back story of the program, hired the kids in violation of child labor laws because adults would have been too expensive. Whenever the kids didn't know an answer, they got slimed.

Nickelodeon executives seemed to understand the semiotics of slime, and the network's creative director mused to the *Wall Street Journal* that the substance "became a symbol of the solidarity of oppressed kids. It's tough to be a kid in the adult world." When Mattel teamed up with Nickelodeon to market the product, they found out how much kids apparently agreed with this assessment.

The righteous indignation symbolized by Slime and confirmed by its huge sales was not to mark the culmination of goop's use as a developmental tool by the youth market. A new Nickelodeon game show called *Double Dare* capitalized on children's love of sticky, gooey messes by allowing them to compete in relay races involving ice cream, eggs, and other viscous substances. The resulting generic mess was termed "gak," and Mattel rushed to create a synthetic equivalent.

Gak's success in the marketplace has been astounding. It has been banned in several public schools—just one testament to its overwhelming popularity. Among Gak's most remarkable properties is that it feels wet, but leaves little wetness on the skin; it has all the joy of moist intimacy with none of the lasting side effects. Bacteriostatic agents have been mixed in to prevent germs from growing within the putty, or spreading from child to child. Designed to imitate the approximate moisture level of an air-dried human tongue, Gak has no purpose other than to feel weird and make "fart" sounds when it is pressed back into its plastic, splat-shaped container.

Toy marketers are particularly impressed by Gak's almost generic iconography. It is the first superhit product spin-off from a kids' TV show that is not based on a character or one of his accessories. The popularity of *Power Rangers* or *X-Men* dolls, vehicles, and weaponry is comprehensible by old-paradigm marketers: Kids identify with the superhero characters, and hope to emulate them vicariously through their action figures, and by using their weapons and possessions.

But Gak is not a character, nor is it used by a character. It's just ooze. This brings us to yet another stunning aspect of kids' experience of evolution and modern intimacy. Just as a teenager experiencing his own sexuality for the first time begins to question or disregard adult authority, rules, and regulations, a society moving into true intimacy must learn to disregard some of its long-maintained cultural barricades. If Al Gore and Newt Gingrich have Internet accounts that display their ASCII text in the same font size as a janitor in Tuskegee, then the relationship of people with their heroes will necessarily change. Idolatry of all kinds will tend to diminish as people begin to value intimacy with equals over identification with a role model. Further, Gak is a toy based in pure play—there is no story, conflict, or metaphorical value. Gak is not *like* goo; it *is* goo. With Gak, you don't pretend to do something else, you play with Gak. Tops and yo-yos were gravity-based toys, whose entertainment value was based in competition, endurance, or technical proficiency. Gak demands no such arc of pleasure or level of skill. It is purely experiential.

Gak says all this? So it seems. Gak initiates the process of social and sexual intimacy as it directs us toward an appreciation of life without role models, expertise, conflicts, stories, or metaphor. You can't "win" at Gak. There's no end to the game. You just put it in its "splat" until next time. In its own small way, Gak represents a non-apocalyptic model of reality, and it comes from the world of children.

While kids didn't invent Gak, they are the ones who made it such a sensation. Toy manufacturers and media executives launch many times more products and programs than are embraced by their target market. When adults take such a scattershot approach to an industry, their chance successes are to be attributed to the particular appetites of children, or the ability of the rare executive to relate to the underlying desires of his audience. One of the very few popular toys that is purchased in roughly equal numbers by both boys and girls, Gak satisfies the three chief requirements for an element from popular kids' culture to be considered an evolutionary indicator: it is loved by kids, hated by their parents, and emblematic of a non-apocalyptic strategy for the future.

We can only hope that kids growing up playing with Gak will understand the world a bit differently than those of us who played Chutes and Ladders. For implicit in the style and context of this new sort of play are many of the leaps in consciousness that will be required for us to navigate our path, as individuals and maybe even a people, through the terrain ahead. Games like Chutes and Ladders were perfect training for a world that understood itself in terms of beginnings, middles, and endings. It was a simple rat race, based on the luck of a dice roll, with occasional random and undeserved grace (ladders) or wrath (chutes). This is the way we, in the West at least, have both rationalized and mythologized our experience of reality. Life is a story, with a beginning, middle, and end. One works toward goals, and God, for no apparent reason, rewards some and not others. The abstract nature of God, fate, redemption, and damnation are all expressed as real-world metaphors: ladders and chutes (or, in the original game, more biblical snakes). The only real law of nature exploited in the game is gravity: you must climb up, but you can always fall down. For the game to end, there must be a winner and a loser.

Well, this game doesn't provide us with any hints for non-apocalyptic thinking. On the contrary, it renders a fairly precise portrait of the more fatalistic profile of Christian thinking at the end of the twentieth century. Everything ends. Life gives in to gravity, unless you are redeemed through some divine intervention. You can't have winners without losers: Salvation of one person defines the damnation of another. The way to understand the seemingly abstract nature of the divine is through simple, physicalized, real-world metaphors, because everything works the same way. Follow the linear path prescribed by the game's rules. There are no choices to make, or strategy to plan, just a die to roll and destiny to unfold. Winning has nothing to do with you.

This way of thinking may have worked during a simpler time, but it is failing us now. It is a worldview that depends on an ending for it to work. Things may get confusing and out of hand, but if you just hang in there the apocalypse will come and you'll be rewarded. Whenever I lecture or go on a book tour one of the most common questions I'm asked about computer culture and the increasing complexity of our media is "Where will it end?" I can only think to respond, "Why does it have to?"

The evolutionary experience of culture, as practiced by kids today, directly contradicts much of traditional New Testament interpretation. It accepts that things keep changing, without a satisfying, determinist ending. It dispenses with storytelling and parable in favor of experiential—or what I'll be calling "recapitulatory"—methods of understanding abstraction or divinity. It refuses to treat the discontinuous as anything but natural: the increasing nonlinearity of our media and popular culture is not a heathen retreat from the dualistic morality of God, but the process by which we learn to accept the very natural, organic, and complex property of life called *chaos*. Worst for the traditionalists who would eliminate evolutionary thinking, sustained chaos refuses us the easy out of an all-encompassing apocalypse. As a kid has trouble imagining himself ever living long enough to make it to adulthood, we have trouble imagining our culture developing past its present, childhood level. Admittedly, an apocalypse sounds better than a stage of civilization-wide adolescence, but this, too, will pass.

What I'll attempt to show in this book is that the more frightening aspects of a non-apocalyptic future are being addressed today, and quite directly, by the most pop-cultural experiences of children and young adults. Whether it's the *Power Rangers* showing us how to accept co-evolution with technology, or a vampire role-playing game calling for us to accept the satanic beast in each of us, these new forms have the ability to assuage our worst fears, confirm our most optimistic scientific theories, and obliterate the religious and cultural absolutism so detrimental to our adaptation to the uncertainty of our times. Within the form and content of kids' favorite shows, games, and social interactions lie the prescriptions for us to cope with cultural change. The oldest of these screenagers have already developed into fledgling filmmakers, computer programmers, and social activists, whose worldviews are beginning to have a felt and, I would argue, positive effect on our cultural values.

So please let us suspend, for the time being, our grown-up function as role models and educators of our nation's youth. Rather than focusing on how we, as adults, should inform our children's activities with educational tidbits for their better development, let's appreciate the natural adaptive skills demonstrated by our kids and look to them for answers to some of our own problems adapting to postmodernity. Kids are our test sample—our advance scouts. They are, already, the thing that we must become.

That is, if in fact we choose to carry on beyond the end of our world, into theirs.

1

THE FALL OF LINEAR THINKING AND THE RISE OF CHAOS

"Skiing is old-fashioned, elitist, and boring—something that your parents do."

Teenage snowboarder, quoted in *The Economist*

Skiing is out. Snowboarding is in. Kids today have no use for the manicured slopes, six-and-a-half-foot ultra-precision skis, aluminum steering poles, or the streamlined jackets of state-of-the-art alpine downhills. Skiing is old-school Europe. Snowboarding is West Coast USA.

Think of the way a trained, adult skier tackles a slope. With nimble finesse, he avoids obstacles in order to glide smoothly down the hill. When all else fails, stem and traverse. Parallel skiing is about controlling one's inevitable descent. The techniques involve steering and, when necessary, slowing down. A snowboarder, in contrast, thrashes his way down the slope.

His feet attached to a single surfboardlike plank, the snow-boarder intentionally seeks out the most dangerous nooks and crannies. No poles; just raw balance. Rough spots are his dominion. "Gap jumping" is among his most important techniques. Snowboarding is about finding breaks in otherwise smooth terrain, and thrashing them for everything they're worth.

The skiers want these kids banned from the slopes. They look and act more like skateboarders than ski pros. They don't understand the etiquette. They don't even take lessons. They wear weird clothes, talk like surfers, and represent a complete break from the time-honored tradition of skiing. And if that weren't enough, they seek out the bumps and avoid the smooth straightaways.

Snowboarders may not consciously understand the intricacies of chaos math, but the evolution of their sport almost perfectly recapitulates the process by which our scientists developed the tools they now use to explore and explain the very roughness of reality. The rest of us had better do some homework to catch up.

Science Hangs Ten

Traditional skiers shun the new kids thrashing down their slopes no more than traditional mathematicians ostracized the first of their colleagues who attempted to use nonlinear chaos equations to calculate their steepness. Discontinuity is intrinsically threatening. Just as there are human genes with no instructions other than to resist mutation, there seem to be human beings with no other programming than to resist cultural change. A certain degree of steadfastness is, no doubt, a healthy thing for both organisms and societies alike. Were there no genes fighting change, a species would not be able to maintain its genetic composition long enough to find a mate, multiply, and pass on any genes at all. If there were no people fighting change, then society would have trouble holding itself together. With no identifiable constants, our world would feel too fluid, too irregular, and too chaotic for any meaningful or survival-enhancing interactions to take place. This is why periods of extreme novelty—a high rate of change—are usually plagued by correspondingly high rates of mental illness, social disorder, cultism, sexual hedonism, and the use of state police or royal guards to keep all this under control.

We can accept a certain amount of linear, or incremental change, without difficulty. Like water being heated, we simply rise a few degrees in temperature, and maybe speed up a little bit. In pre-Renaissance Europe, as the Church grew in power and its models for the world grew more developed and elaborate, there was no major social upheaval. But fundamental, or categorical change presents a very different set of problems. When Copernicus declared that we were misunderstanding our reality, there was crisis. Society, like steadily heated water, has its boiling points. After a long, linear climb accompanied by an increase in motion and kinetic energy the system can withstand no more. The single option remaining is to break from linear intensification and change state.

A societal shift of this magnitude doesn't happen on a whim. People only accept a change of state out of necessity. It happens when the way we've been understanding the world so contradicts our felt experience that it stops serving any useful purpose. A model of reality can be thought of as working when it has the ability to explain the present or predict the future with some degree of success. It reassures us that everything really is all right and proceeding on schedule, according to plan. We would rather ignore or discount any evidence that negates our model, if at all possible, than change our picture of reality. This is why we kill people like Socrates, Giordano Bruno, and Copernicus before we allow their ideas to dismantle our models of the way things work. We eliminate these

17

people not because we are ignorant, evil fools, but because we can only accept so much change at a time.

The Copernican revolution, once accepted, amounted to a full-scale paradigm shift. We used to believe that the earth was the center of a living universe, kept in motion by God. Mankind and earthly affairs were a microcosm of God and heavenly affairs. The universe was understood as a series of concentric spheres, with our planet and activities at the center of it all. To accept what Copernicus was saying back in the sixteenth century—that the earth wasn't really so special—required a pretty substantial shift.

What eventually resulted was the scientific model we have been using through modern times. Now, scientifically speaking, we only believe in what we see or experience. God may have created the universe and even set it into motion, but since then events have been determined by mechanical forces and entropic law. The world is not a living thing with divine purpose, but a dead one, in an existential void. Religion managed to survive only by separating itself from science. In a kind of conceptual apartheid, there was (and still is, to some extent) strict and enforced separation of the realms of science and spirituality. The Church would handle God, consciousness, and the spirit, while the scientists handled physical reality, the motion of objects, and history—as long as they didn't go as far back as Creation.

This simple arrangement brought us from Descartes and Newton right through to Bohr and Einstein. But it also imprisoned our sciences in so-called rational thought, linear thinking, and empirical, repeatable evidence, rendering them defenseless against the latest incursions of reality on the human experience. Meanwhile, the separation of science from religion divorced our spiritual sensibilities from real life. By ignoring at least half of human experience, science, as well as the technologies and culture it spawned, tended to resist nature rather than promote it. Whether it was electric lights letting us defy the natural darkness of night or the theory of behavioral psychology letting us deny the ambiguity of a force like "love," science helped us reduce a messy reality into simpler and neater sets of boxes. This reductionist thought literally reduces the complex into manageable, if artificial, components. For many applications—like charting the forces acting on a wheel or space shuttle—this works splendidly. For many more other real-life events, this application of technology fails miserably.

It's the motivation behind reductionism we have to understand before we can be liberated, as our kids have been, from its stultifying side effects. The purpose of reductionism is to contain and manage nature. To reduce something to a category puts it in a convenient, if unreal, box. Its qualities and its behavior can be talked about, because it is officially limited. For our mathematics to work, we ignore the inconsistent roughness of reality and smooth out surfaces in order to make them fit into our equations. For example, mathematicians have great equations for calculating the surface area of, say, a perfect cone. But how do you calculate the surface area of a mountain? There are innumerable irregularities. Do you measure around each boulder? Each rock? Each nook and cranny in each rock? Each grain of sand? The branch of math called calculus was invented to help us cope with the problematic irregularities of real life, by reducing them to flatter approximations of their former selves. But this is cheating. Did math really get all the way to the twentieth century without us being able to calculate the surface area of a real-life surface, or the coastline of an island? No wonder kids hate math. It doesn't work.

Worse, we've applied reductionist thinking to so many real-world problems that we've dangerously reduced many of our real world's properties. Deforestation renders extinct about

100 of the planet's plant species each day, reducing plant bio-diversity by up to 11 percent each decade.* Our disease-based medical model, which reduces germs to simple enemies, failed to take into account that there are many more of them than there are of us. Each new, painfully researched antibiotic is outsmarted by a few billion bacteria, which quickly mutate into resistant strains. Our agricultural practices, which use chemicals to kill insects rather than nutrients to strengthen plants, have developed a population of extraordinarily tough pests and a crop of weak, drug-dependent crops. By attempt-ing to stamp out the inconveniences of nature, we have made ourselves all the more vulnerable to them.

Scientists, mathematicians, and children have begun to face the fact that facts must be faced, not reduced to tempo-rary and deceptively quiet obscurity. This new willingness was embraced first by the physicists, who admitted (or discovered) through the "uncertainty principle" that nothing can be observed without being changed through the act of observa-tion. The tiny particles of their experiments changed their motions, depending on where the scientists looked for them. This wreaked havoc on a scientific model based solely on the sanctity of observed phenomenon.

While most of us don't believe that this obscure principle from the thought experiments of quantum physics has any direct influence on our lives, we are accepting more and more of its implications almost every day. Although a thoroughbred scientist would shudder at the comparison, consider how a media event like the O. J. Simpson debacle reflects the very same principle. As the Los Angeles police pursued O.J. in his Ford Bronco, millions of people watched on their television sets. Many Los Angeles viewers realized that O.J. was about to drive by their homes and ran outside to wave him on. As they rose from their couches, leaving their TV sets turned on, these

*National Wildlife Federation, World Wide Web site, 1995.

people literally walked out onto their TV screens. They became the very story they had been observing, adding to the event. The story itself changed as the cameras and commentators began to focus on the growing crowds of spectators, which in turn led to more people running outside to join in the event. As an observing population, like the scientists, we saw ourselves under our own magnifying glasses. The observers and the observed became indistinguishable.

We are growing to accept the implications of our paradigm-bending sciences not because they happen to be right, but because they more accurately reflect our current cultural experience. Contrary to popular academic belief, the theories of high science do not trickle down to culture at large. The needs of culture at large trickle up to the academics, who scurry to come up with models that answer our experience. They are, after all, members of popular culture, too.

There is no record of a great public outcry over the way mathematicians had been measuring the surface area of mountains. Nor was anyone except the executives at IBM worried about the seemingly random line interference on their data transmissions (the intermittent but untraceable lapses in the clear telephone connections they depended on for high-speed data transfer). But Benoit Mandelbrot's work with discontinuous equations in the 1960s did come precisely at the time when our culture was attempting to cope with the unnerving discontinuity of the modern age. A telephone ring can bring a family tragedy to the dinner table, a television can instantly report and replay frame-by-frame the assassination of a president, and the press of a single button can wipe out an entire city instantly. The increasing fragility of continuity in the modern experience made it incumbent upon us to find another way to understand the flow of events in our lives.

Mandelbrot figured out that the way to cope with discontinuity was to embrace it, not shy from it. He used what are called nonlinear equations—ones that, rather than providing an immediate answer, keep taking the answer and plugging it back into the equation, thousands upon thousands of times. Linear math would say that x + 1 = y. If you decide that x equals about 3, you plug it in and get 3 + 1 = 4. You got your answer. Nonlinear equations don't end quite so quickly. You take your answer, in this case 4, and plug it back in as a new x. Now x equals 4, and 4 + 1 is 5. Now x equals 5, and so on. In this vastly oversimplified case, x will slowly move toward infinity. Other cases might bring x to zero, a constant, or nowhere at all—just some wavering value. When Mandelbrot graphed these nonlinear equations—which can only be done with the aid of computers capable of churning through these nearly infinite loops—he got fractals: those brilliant, paisley designs that caused such a stir in the early nineties psyche-delic dance culture, "rave."

One amazing thing about fractals is how organic they look. A fractal image can depict the intricacy of a coral reef, a fern plant, a canyon, or a snowflake. Fractal equations certainly draw much more accurate representations of our world than the smoothed-out cones and cylinders of traditional geometry. The other astounding property of fractals is that, despite the fact that they are based on discontinuity, they exhibit extreme "self-similarity"—an almost impossibly intricate set of self-mirroring patterns. If you identify one of the many shapes or surfaces in a fractal picture, and zoom deep into the nooks and crannies within that shape, you will find smaller, almost identical versions of that same shape. If you magnify the tiny nooks and crannies within the tiny version, you will again find more self-similar shapes.

Well, surprise surprise, it turns out the rough-hewn edges of our very irregular natural world also exhibit this property of self-similarity. The surface of a mountain, viewed from afar, will have characteristic irregularity—a certain set of shapes defining its bumps and valleys. If you examine the surfaces of the rocks making up those bumps and valleys, they will have similar nooks and crannies, and so on. This self-similarity is not as perfect as traditional scientists would like: the self-similar shapes are never exactly the same as one another, but they are close enough for us to recognize them as related. We have to accept our interpretative ability as real.

Mandelbrot used repetitive loops to magnify the irregularities of the world, rather than smooth them out. Take them on their own terms, and see where *they* take *you*. He found that what appear to be senselessly chaotic irregularities actually have an underlying order to them. The random blips and squeaks of line interference at IBM, while inexplicable in normal, sequential math, nonetheless had shapes to themselves. They were fractal in essence, and each burst of interference, when broken down, had tiny bursts of interference within it. The larger bursts were spaced apart at intervals self-similar to the space between the tiny bursts within them. And within these tiny bursts were other, tinier bursts, also occurring in an irregular pattern similar to the pattern of interference on each "higher" level. Mandelbrot had found the pattern, or order, underlying the chaos.

By focusing on discontinuity, rather than avoiding it, we can come to understand its nature. Fractals are now used to help reckon with what were formerly incomprehensibly complex, chaotic systems like the weather, stock market, and even the level of violence or passivity in a population. Once we understand any pattern of the system's irregular behavior on one of its fractal levels, we can begin to recognize this same pattern on any of its others.

This leads to an appreciation of discontinuity as a source of reassurance, and a wholly new understanding of what has come to be known as chaos. Chaos is not mere disorder—it is the deeper order within apparently random, nonlinear systems. Chaos is the character of discontinuity.

Simple, linear systems do not have these same, lifelike properties. It's like the difference between a line dance and a rave. The line dance is predictable, and the number of possible interactions is fairly limited. At a rave, where individuals dance freestyle according to their momentary whims and impulses, almost anything is possible—although, deep down, the mood and character of the people dancing may reveal certain characteristic gestures, clumps of people, or overall movements. Think of a dripping faucet. When the drops are falling one at a time, or even as fast as a trickle, the motion of each drop is pretty predictable. When the faucet is opened so wide that the water attains turbulence, then the motion takes on a whole new set of properties—what is now called chaos. When a system has gotten so complex that it fails to be predictable by linear means but exhibits the emergent behaviors of chaos, it is called a dynamical system.

Dynamical systems such as the weather, an organism, or a school of fish are more, not less, interrelated and interdependent than a neat, conceptual hierarchy like a classroom, factory, or machine. The tiniest change within the tiniest detail of a dynamical system can implicate huge changes on a higher fractal order. When one of the tiny microbes in a coral reef decides to begin the reproductive cycle, it releases an enzyme and changes color. Within hours, hundreds of miles of adjacent reef, triggered by the first tiny microbe's enzyme, also change color as they move into the reproductive mode. But this is not a simple cause-and-effect linear relationship. The tiny microbe would be called a "high leverage point" in the dynamical system. Were the other microbes not also ready to begin the reproductive cycle, the first microbe would have just changed color alone. But because its timing and position are perfectly placed, the microbe serves as the leverage point for the rest of the reef to activate.

This led to the well-known but little understood notion that a butterfly flapping its wings in Brazil can lead to a hurricane in New York. The weather is such a dynamically interdependent system that one tiny event on an extremely "small" level of the fractal can provoke a systemwide change. As I attempted to show in my book *Media Virus!*, a popular cultural example of such a high-leveraged butterfly's wing would be the Rodney King tape. One homemade videotape, recorded in the right place at the right time and then iterated through the dynamical system of world media, ultimately provoked riots in a dozen American cities.

This is why chaos can be so disheartening to those who would control their lives or even the rest of ours. Change can come from almost anywhere, and only the change that the entire system is "ready" for can be fully triggered. Chaos empowers the tiny, while acting as nature's truth serum. No wonder kids love the stuff.

Now Boarding

Anyone who has ever tried to build a sandcastle on the beach or a home in Malibu should probably not expect it to be around too long. The ocean is a dynamical system that shows little respect for the structures we impose on or near it. Surfers, among the first true children of chaos, have long cherished the chaotic nature of the waves they ride. Their intimate relationship with a dynamical system as complex as the ocean led to a roll-with-the-punches lifestyle and postlinear spiritual outlook. Though a bit incomplete as an entire societal philosophy, surf culture and the subcultures it spawned are among the first to call chaos home.

Go to any surf shop or leaf through the pages of a surf magazine and you will immediately be struck by the graphic images on the T-shirts, surfboards, and decals. Yin-yangs, detailed pictures of waves, and fractal images abound. This is a culture based on the wave pattern of the ocean, and the sensibility is that learning how to navigate that pattern is both an enterprise worthy of lifelong dedication and an exquisite rush.

The Taoist yin-yang symbol is a natural emblem for the surfer. Not only does an ocean wave, when viewed from the side or the inside (as only surfers get to see it) physically resemble the swirling yin-yang image, but the cyclical course of nature that the symbol suggests serves as the underlying surf philosophy. There's always another break, another wave, another tide, another day, another season. A surfer should be equally "stoked" by a terrific wipe-out as by a brilliant ride, and fellow surfers should respect both outcomes equally. No judgment. The balance of opposites represented by the yin-yang is also incorporated into this sport-that's-not-a-sport. The "professional surfer"—almost an oxymoron—competes vigorously for top honors and huge purses at a heat, but does so by relaxing into a zenlike appreciation of the afternoon's wave phenomena. Like tai chi chuan, this is a moving medita-tion dependent on the practitioner's ability to marry intention with sensitivity—will with passivity. In this sense, the expert strikes a perfect balance between ego and universe, and he intentionally immerses himself into an overwhelmingly chaotic system in order to do so. Pretty trippy.

The organization of a surf meet tests not only a surfer's wave-riding skill, but his ability to master an entire wave sys-tem. The competitors paddle themselves out from the shore, and then pick which waves to ride in a given time period. While competitors are judged for their ability to make some-thing out of nothing, if you pick the wrong, wimpy wave you won't have as good or as long a ride as someone who has picked a wave that turns out to be a flawless, epic tube. Getting a sense of the entire pattern, and being able to predict which waves will provide a good ride, is easily as important as how well the surfer negotiates the individual wave.

A surfer at a particular beach who has an uncanny ability to pick the right waves and then to know exactly how to surf them is said to "own" that section of shore. Usually this will be a beach near his hometown, where the set of waves is familiar to him—his own neighborhood of tidal and wave patterns, his own piece of the ocean. A fractal, too, though as infinite in scope as the wave patterns of an entire ocean, is experienced by mathematicians as having neighborhoods, identifiable by a set of coordinates and a level of magnification. One neighborhood of a fractal may have sharp, discontinuous dots, while another may have long, weblike tendrils. A mathematician might be drawn to a fractal neighborhood for its ability to explain a natural or theoretical phenomenon, or he may just be drawn to its beauty or self-similar quality. Different mathematicians become experts of different fractal neighborhoods. It is in coming to understand and recognize the similarities among different dynamic neighborhoods, as well as the qualities within particular neighborhoods, that surfer and mathematician alike learn to cope with chaos.

A surfer drops out of mainstream life to hang out with his colleagues in a mutually supportive surf environment, much as a dedicated mathematician locks himself away with his colleagues at the university. Although the surfer's outward appearance is quintessentially "slacker" (no more disheveled than many a math professor's, actually), he is on a quest for a Holy Grail of sorts. Just as the mathematician searches for the ultimate equation to explain the universe, the surfer waits and watches for the perfect wave, accepting finally, as one champion wistfully mused to a surfing magazine, that "most perfect waves in the world go unridden." Similarly, mathematician Paul Erdos believes that the greatest math formulas are already in existence, and that geniuses like Einstein simply discover something "that's in the book" already. Both math prof and surf dude will be the first to admit that the only thing they can do is be at the right place at the right time, ready to recognize and exploit the opportunity if it arises.

To the experienced surfer, "real" life works the same way as life on the waves. As one sixty-year-old surfer explained to *Surfer* magazine, "Name something that isn't surfing. In the

New York Stock Exchange, you check it out, you pull in, and you try to figure out when to pull out safely."*

Like much of the activity in the culture of chaos, the experience of surfing is often compared to birth or evolution. Surfers talk about staying in the tube for as long as possible before being ejected or "spit out." The visual quality of passing through an unfolding tunnel of water is like the tunneling imagery of psychedelic movies and rave videos, which are themselves often explained as reminiscent of the journey down the birth canal. Hang Ten Sportswear's 1995 ad campaign employed a slogan evoking both evolutionary and earth imagery—"All life comes from the sea. Get back to the womb"—in neotribal typeface superimposed over a surfer emerging from a giant pink tunnel wave. The company's tag line, appearing right under its logo, is a call for intimacy more blatant than the slime toy Gak: "Stay wet."

This is where, for some, surfing falls short of offering a complete strategy for life in the chaotic future. With its back-to-nature tendencies, surfing often seems like a call to go backward in time, toward a simpler way of living. While the insights it offers us into the practical and spiritual navigation of dynamic systems is invaluable, the surfing life seems dependent on a certain amount of social isolation and yields, in the long run, an antievolutionary solution to social and biological woes. Because, as any good surfer should know, there is no going back.

Urban Waveforms

Skateboarders based much of their culture and ideology on the surf ethic, but instead of riding the chaotic waveforms of the ocean, they content themselves with the discontinuous

*Tom Morey, quoted in *Surfer* 36, no. 4 (1995).

complexity of the man-made cityscape. Curbs, banisters, fountains, and staircases make up the skateboarder's terrain, and for him to apply the same laws of dynamical systems to the urban landscape requires that he accept the city and modernity itself as an expression of chaos. You just need wheels to surf it.

Like surfing, skateboarding depends on the ability to make the most of a particular set of surfaces. Tricks develop around obstacles and are based on transforming an impediment into an opportunity for a flourish like a "switchstance kickflip nose wheelie," or a "frontside heelflip noseslide." Any skateboarder worthy of his board can tell you exactly what those moves would involve, as well as which kinds of surfaces—like a park bench or wheelchair access ramp—would be best suited to their execution. Other boarders fit warehouses with large, parabola-shaped "vert ramps" on which they practice flips and conduct tournaments. Some commercial facilities are also available where skateboarders pay a small fee to practice on state-of-the-art slopes. Vert ramps have declined in popularity, however, because they don't truly fit the skateboard mentality. Why construct a perfectly curvilinear skateboarding simulacrum when the whole point is to thrash through, around, and over the impediments and imperfections the real world offers? While a few skaters, mostly those over thirty, remain dedicated to their prefab skateboard environments, most kids prefer to take it to the streets.

This is why skateboarding may have as much in common with hip-hop as it does with surfing. It is a city sport, and finds its ethic and aesthetic in blacktop culture. The fashion is sneakers, baggy T-shirts, loose pants worn low on the hips, and backward caps. Pot is everywhere; skateboard magazines are filled with articles on scoring buds, bong-athons, and ratings of new cannabis strains. The music is metal or rap, depending on the neighborhood. Skateboarders ride in small posses, each with its own name and graffiti tag.

Skateboarding and graffiti are inextricably linked. Skaters often travel with markers in their back pockets, and their sport provides a convenient and speedy getaway. Skating and graffiti both use the cityscape as canvas and proving ground. Conquering a tunnel or overpass could mean navigating it on your stick, successfully spray-painting your tag, or both. Graffiti tags are not simply random acts of vandalism. Like skateboarding a local nook or cranny, tagging a landmark gives it personal, totemic value. A skateboarder and his gang can tag and thus claim thousands of locations throughout their neighborhood, transforming the adult, man-made, but ultimately natural urban landscape into a series of power spots. As in ancient Native American tradition, astrology, or even modern mathematics, the pursuit is to turn an otherwise random, patternless complex of trees, rocks, stars, or alleyways into a personally understood tapestry, with internal reason and exploitable power. The city becomes a series of overlapping grids, corresponding to the various local skateboard posses and their individual members.

It's as if the kids are developing a series of resonant fields, each impressing a set of values—however topically inane—on an otherwise completely meaningless concrete wasteland. Taken alone, and certainly as viewed by a property owner, a graffiti tag may appear to be a single, discontinuous blot on a formerly smooth wall. But when understood as one of many such tags thoughout a city, it gains an almost metacontinuity, from tag to tag. The ability to recognize such overlays of continuity is one of the key skills for a denizen of chaos.

This quasi-spiritual quality of graffiti has permeated skateboard culture, reaching a near obsession with stickers, logos, caps, T-shirts, and decals. The skateboard itself is a work-in-progress, not just in terms of its ever-changing hardware fittings, but in its overall appearance, which is constantly being modified by new store-bought logos, and handpainted tags. And weird ones: alien images, UFOs, a version of the

Christian fish bumper plate, with little legs sticking out and the word *Darwin*. The most popular new tags today concern the core themes of coping with evolution, the future, and chaos.

Skateboarders' clothing is also extremely tag-conscious, even more conspicuously and meaningfully than in the conventional fashion world. Professional teams are organized under specific labels, and a few skaters have even allowed themselves to be tattooed with their company's logo. Often, rather than just relating the name of the designer, the label promotes a social agenda, however simplistic. Wearing the "fuct" label is a statement in itself, and the company's slogan, in blaring self-consciousness, proclaims "profanity is profit." So immersed are skateboarders in the graffiti aesthetic that they not only spread their own tags throughout a neighborhood, but act as tiny mobile billboards themselves within larger frameworks upon frameworks of tags and billboards, all exhibiting the self-similarity of a fractal.

In several ways, then, the activities of the skateboarding graffiti artist are analogous to those of the chaos mathematician. The kid needs skill, a skateboard, and the ability to understand the patterning of the cityscape in order to participate in the conversation of tags and glyphs. His physical maneuvers are captured in frame-by-frame photography and deconstructed in skateboarder magazines. His posse gives him camaraderie and reinforcement, and a corporate sponsor, if he can get one, trades him money for labeling his efforts with its tag. The mathematician needs skill and computers to get himself transported to a particular location within a fractal neighborhood, enough familiarity with that neighborhood to recognize and isolate distinguishing "taggable" features, and a posse of peers to recognize his achievements, step by step, in juried publications. If he is successful, he will be offered sponsorship by a university that will attempt to claim him as a member of its tenured, branded faculty.

For however advanced and nonlinear skateboarding may be, it is no less motivated by ego than is mathematics or any human endeavor. Most skateboarders you talk to will have little to say about fractals and resonating fields. These subtler aspects of the sport go on for the most part on a subconscious or, as I'd argue, on an instinctual level. This is simply what human beings do to feel comfortable in a seemingly discontinuous reality: they recognize or create patterns encompassing one another. But, almost invariably, individual ego is the baser engine driving these high-minded, communitarian projects.

An average skateboarder's conscious motivation is not to forge a new postmodern civilization, but to take the most daring personal risks possible. Traveling through Manhattan by holding on to bumpers of cars and buses is one way to prove your merit; grabbing police cars that have their sirens on confer highest status, with firetrucks and ambulances close behind. Again, the thrill is always to convert existing urban geography or activity into pure entertainment, but the gratification is largely self-aggrandizing. The greater the personal risk, the greater the personal achievement. Young white kids are proud to skate in black ghettos late at night. Others leap across highways or attempt stunts that, if unsuccessful, would land them in the middle of traffic.

Many fan magazine profiles of famous skateboarders feature long narratives and detailed photographs of bloody injuries and broken bones. Sometimes such stories describe, with suitable awe, how the injured skater refused medical treatment, leading to even greater risk. "Another couple of hours," one skater explains about his delayed decision to get medical attention, "and the doctor says I would have lost my leg or died." Not unaware of this fascination with peril, skateboard and fashion advertisers appeal directly to their customers' blood lust. One ad for sneakers shows a boy lying dead in a puddle of blood on the pavement beneath an overpass. He has a morgue tag tied to his toe, necessitating a cross-sectional cut through the shoe being advertised. A successful skateboarder, all bruises and stitches, whether alive or dead, is the chaos kid's equivalent of a war hero.

Unlike surfing, which has a legacy of elders and styles, skating heroes are nearly all kids in their teens or early twenties. Their stars are their peers, making a high-profile skateboard career seem in reach of any daring enough teenager. The hero is elevated yet one of us at the same time. The cash prizes, posters, calendars, hotel suites, and beauty queens are the rewards not for some remote superhuman star but for some kid from Orange County. That's half the rush. They're putting up some kid in sneakers and jeans at the Ritz-Carlton! We've cracked the system.

The kids who do make it to the top sometimes half-ironically adopt the attitude of a martial arts master. One champion skateboarder calls his style "The Way of the Intercepting Fist." Another, in an ad for his sponsor, is pictured with the caption: "Mike Cao: Master of the Wing Ding Style." Skateboard magazines are filled with ads for videos of Hong Kong action movies and kung fu video games, both traditions based on characters who learn, evolve, and perfect their personal versions of particular martial arts forms. Without an immediate lineage of their own, skateboarders have adapted one from media history— maybe the only one they've ever heard of. They haven't really incorporated a lineage into their sport, but rather the *idea* of lineage. Like the flying saucer skateboard stickers that refer to our sci-fi or evolutionary future, the many references to ancient martial arts systems recontextualize skateboarding as part of a larger cultural progression. Discipline, form, mental clarity, spiritual purity, and all the other qualities associated with kung fu mastery are now also components of the skateboarder's life strategy.

This may seem somewhat contradictory at first. If they are so liberated from tradition and linearity, why do skateboarders cling to imagery from the ancient Shaolin temples? If they

are beyond the notion of ego-based, hierarchical constructions, then why are they so dedicated to daredevil antics and
personal promotion? This is a core paradox of chaotic living,
but an altogether reconcilable one.

First, the incorporation of apparently linear schema into
this otherwise discontinuous sport and culture is a bit decep-
tive. It is the *notion* of lineage that the skateboarders admire.
Its incorporation amounts to a nod at history; tradition has
been reduced to an icon—one among many in the pantheon of
glyphs. If anything, this supports a more nonlinear appreciation of historical time, where useful moments and strategies—
even the stategy of linearity itself—can be absorbed on an as-
needed basis, divorced of their original contexts. To earn the
tattoo of a Shaolin priest traditionally required years of dedicated practice and the carrying of millions of buckets up tem-
ple hills. Now it's a self-conscious fashion choice.

As for the seeming incompatibility of ego and group consciousness, this is based on the faulty, old-paradigm view that
individual expression and group cohesion are mutually exclusive. They are not. What chaotic dynamics demonstrate is
that, left to their own devices, the elements of a complex system will maintain great order and stability. It doesn't look like
the order we're used to, with straight lines and readily identifiable categories; it instead reflects its order through the less
tangible and less exact qualities of self-similarity, interdependency, and high leverage points.

Skateboarders demonstrate extremely aggressive individu-
alism at the same time that they are being actively co-opted by
corporate sponsors. Meanwhile they experience their primary
allegiance to their local posse of amateurs or team of fellow
professionals. As we'll see again and again in the activities of
the children of chaos, the aggressive expression of individual
and local mentality almost always supports the embrace of a
larger fractal or community entity, even if that entity is corporate in nature. When a group of skaters brand themselves with

34

the name of their sponsor, they are engaged in a pagan, neo-primitive ritual that would make most of the company's share-holders cringe to witness it directly. Tattooing and piercing by teenage boys is seen by most adults as antisocial, countercultural self-mutilation. The kids see it as willful self-modification. Ironically, it is also a form of advertising.

So isn't it possible for everyone to be happy? The kids get to scandalize their parents by breaking family traditions and to establish personal allegiances with their friends, while the companies get to spread their logos and sell their products. Through their example, the rest of us get a valuable lesson in how to recognize the order in chaos and dispense with the obsolete belief that aggressive individual expression in any way sacrifices the ability of such individuals to cooperate within an overall system.

If all this seems like a lot to attribute to the sport of skateboarding, you won't even want to consider snowboarding, where the participants claim a full awareness of how their pastime demonstrates nearly all these chaotic and evolutionary principles. Take the Hang Ten ad for snowboarding clothes, which mutates their own surf ad. "All life comes from the sea," it begins, in a word-for-word duplication of the original before veering off into "Then it heads for the hills." Contradicting the retroevolution of the previous ad, which called for surfers to "get back to the womb," the snowboarding ad quite directly appeals to the snowboarders' contention that they have spurred the evolution of the boarding sports forward. Snowboarding is a self-consciously evolutionary designer sport.

Snowboarding is both more overtly natural and patently artificial than skateboarding. While the boarders go up to the mountains and back to nature, they race on man-made slopes at cordoned-off resorts. Almost no one snowboards to get to school or deliver newspapers. You can't snowboard through the neighborhood. You have to make a conscious effort to travel to a slope, purchase a lift ticket, ride in a conveyance of one sort or another to the top of the hill, and then snowboard down. For however economically elitist this makes the sport, it also affords snowboarding an intentionality and contemplativeness that skateboarding and, to a lesser extent, surfing lack.

While surf music is lazy beach boy and skateboarding's is street rap, snowboarding is heavily associated with the rave. Rave parties, where thousands of kids dance to digital music, are planned as consciousness-altering events. The psychedelic drugs, music, and lights are designed to put everyone into a group trance. By the end of the evening (which means dawn) the kids hope to experience themselves and one another as parts of a single, metaorganism. It's both futuristic and intensely tribal, making use of technology to promote deeply spiritual agendas. But snowboarding and raving have more than their music in common.

The overlap between snowboard and rave cultures includes fashion, outlook, philosophy, and people. San Francisco is the center of both of these worlds, because the city offers ravers a supportive, free-thinking psychedelic and computer community, while giving snowboarders all this plus the nearby slopes of Lake Tahoe. Clothing designers who used to make boyish, colorfully clownlike ravewear quickly adapted to the needs of snowboarding. As a result, snowboard style has been infused with the values of the rave: unity, tolerance, individual expression, and global evolution. That snowboarders dress like rave kids, then, should come as no surprise: They are the very children of the rave, who have adapted their sport to the goal of designer consciousness.

The owners of the Yang clothing line, for example, are prominent figures in the San Francisco rave and psychedelic communities (their motto is "Clothes for your Trip") and served as consultants to Oliver Stone on matters psycho-pharmacological for his TV miniseries *Wild Palms*. Many of the graphic designers responsible for snowboarding magazines, advertisements, and logos also create rave promotional flyers, album covers, and videos. As we'll see when we consider rave more fully, this aesthetic involves cultural "sampling"—the selection and recombination of sounds, images, and ideas from throughout history. This is a quite conscious effort to reach an eternal, timeless moment of bliss by recombining, overlapping, and juxtaposing imagery from multiple periods within a single beat of music or frame of film.

Snowboarding is fraught with such contrasting influences. It is a back-to-nature sport, yet highly technological. It promotes the peace-love rave ethic while praising the testosterone-induced antics of its adolescent stars. It has its own dual lineage, coming from the surf/skateboard tradition on one side and San Francisco rave culture on the other. It requires a pretty substantial cash investment from its participants yet appeals to kids who like to think of themselves as outside the mainstream rat-race culture. Meanwhile, these same kids get picked up by corporate sponsors and then make no secret of their financial success.

Snowboard marketing reflects this union of opposites. Ads combine characters from Japanese *animé* cartoons with photos of snowcapped mountaintops; scientists in lab coats working on a physics equation with a comic-strip dream balloon containing a brilliant photo of a boy flying on a snowboard; or plastic *Star Wars* figures framed within a snowboard clothing label. Another shows adult comics–style cyber-beatniks smoking cigarettes under the caption "We are the people our parents warned us against: Shuvit snowboards" followed by a lovey-dovey rave slogan, "Relax, live life, love" in puffy skateboard-style graffiti. One magazine layout on summer snowboard activities depicted a maze leading to Uma Thurman—herself a movie star, daughter of a Buddhist scholar, godchild to Timothy Leary, and, as we learn here, snowboarding fan.

Snowboarding is also about jumping gaps. This is apparently an inexact, instinctual process. As one technique article by a snowboarding champion offers in all seriousness, "It's usually always better to overshoot a gap than underestimate it." Usually always? Exactly. The language betrays the improvisational style of the entire sport. Just as the snowboarder/writer increases his degree of certainty during the very act of writing his sentence, the snowboarder pushes himself harder during the jump, for it is, according to the experts, usually always better to do that than to fall short.

Snowboarding is an intentional exercise in relaxing into chaos. The precision control of the skier is gone. Because of their pole-less, look-Mom-no-hands technique, snowboarders have a very hard time steering, and the skiers with whom they share the slopes complain about frequent collisions and runaway snowboards. Worse, skiers say, is the awful sound of a snowboarder gaining from behind them on the slope. Snowboarders crunch against the grain of the snow, making terrific noises that send traditional skiers into panic—the same sort of queasy panic parents feel watching their kids play Power Rangers in the living room, or any middle-aged person feels when he realizes his Social Security checks will be depending on the likes of, well, snowboarders. We may not like who's following us down the slopes, but they're gaining on us.

Unlike parallel skiing, snowboarding is a sport designed to give its participants the opportunity to test their reactions to gaps of many kinds. Whether it's the conceptual gaps in a piecemeal culture of surf, rave, and science or the physical gaps between mounds of snow, the coping strategy is the same: "Usually always overshoot" and definitely always relax and have a good time. This is only as meaningless as Beavis and Butt-head or Gak—which means not at all if we grant any significance to the moment-to-moment experience of these kids.

The art of coping with a high-tech, nonlinear culture is the art of surfing, skateboarding, and probably most of all, snow-boarding. For where surfing is a negotiation with the dynamic waveforms of nature, and skateboarding is a negotiation with the cityscape, snowboarding is an immersion in designer discontinuity.

Channel Surfing

Just like ocean surfing, the habitual channel surfing of our TV-fixated youth is as lamented by parents as it is valuable to us all as an example of thriving on chaos. Just like the ocean or the man-made cityscape, the modern mediaspace, too, is a chaotic system, and subject to the same laws of dynamics. Better yet, for this same reason the media is surfable. This sport of couch potatoes offers both a lesson in coping with discontinuity and a possible challenge to the status-quo-promoting (and evolution-resisting) linearity of our traditional media.

The old style of viewing television involved commitment. We would decide which program we were going to watch, turn the dial on the television set to the network (one of three) for the evening, and passively absorb our programming. And it wasn't called "programming" for nothing. We were a captive audience, relatively unable or at least uninspired to make choices while we watched. We "behaved" like good little school children, and when Walter Cronkite ended his television news broadcasts each evening with a reassuring "And that's the way it is," we had no reason to question him. The TV was our parent and teacher. To shut it off or change the channel in the middle of a program was a deliberate statement of dissatisfaction.

With the knowledge that they had a captive audience, programmers and their sponsors were free to persuade us by using linear arguments—by telling stories. Time was on their

39

side, and the instantaneous visual language of the image was still only in its infancy. The highly developed but somewhat infantilizing techniques of public relations were perfected during this era of linear programming, and they all depended on narrative storytelling. Like fables, PR campaigns follow a line of reasoning: (1) here is something you care about; (2) here is a terrible threat to that something; and (3) here is the way to annihilate that threat. We came to America and, with God's help, created a paradise, but the evil, atheist Russians decided to take over the world. They are about to invade—what can we do? Build weapons now.

As viewers, we grew used to "staying tuned" for the answers to the world's problems, and by the time the broadcast was over, we knew how to feel, buy, fight, or vote. As children of linear reasoning, we valued the attention span and the ability to draw the correct conclusions from the data presented to us. Of course, we had the announcer to tell us whether our conclusions were in fact correct, and quiz shows with which to practice. It never occurred to us that the shows could be rigged or that the announcer could be just plain wrong.

The introduction of the remote control altered our relationship to the tube forever. We no longer had to make the grand gesture of walking up to the television set and physically rotating a dial in the real, material world in order to change the picture on the screen. (Television viewers of the 1940s and 1950s were further inhibited by their memories of the days of radio, where tuning in a new station was an act that required even greater effort.) Thanks to the remote, a simple channel change no longer signifies rebellion. A tiny motion of the viewer's finger wipes Cronkite's successor's image from the screen, along with his message. The linear argument is broken, and a gap is introduced—for the viewer's weapon against programming is discontinuity.

Dramatic television was even more dependent on linearity

for its effectiveness as a persuader and marketer. As Aristotle well understood, drama works by creating and then releasing tension. The structure of such a story is like an arc, rising to a turning point, and then falling again. A successful teleplay will follow a formula analogous to the newscast: (1) create a character we like; (2) put that character in peril; and (3) rescue him. The object of the game, for the programmer, is to generate as much tension as possible—make the situation so horrifically unresolvable that the audience begs for relief. To raise tension, the television dramatist needs to create a pressure cooker. Within this closed system, the captivated audience is led up a linear thread and eventually over the arc of the story. (The occasional commercial break, though a potential lapse in continuity, is bridged as best as possible with an interim cliffhanger. Ideally, the cliffhanger leaves the viewer in wide-eyed, anxious suspension, all the more ready to absorb the advertising content.)

In linear drama or news, the audience is only willing to put up with the increase in tension if it knows it is going to be rewarded with a satisfactory conclusion. With the knowledge that they are going to be rescued, viewers allow themselves to be led into great confusion. The more desperate the story gets, the more preposterous the conclusion is allowed to get. If we are driven into tremendous tension, we will accept almost any relief. If the entire Athenian civilization is at risk, then Athena herself can fly in, deus ex machina style, and save the day. If innocent babies are being yanked from their incubators and allowed to die, then an American invasion of Iraq feels justified. We have been brought into tension, and we demand relief. The perpetrators must be identified, the guilty punished, and the innocent somehow bettered by the whole experience. From Aesop's fables to *Dragnet*, whoever gave us relief from the tension was allowed to tell us the moral of the story.

Television, for a time, perpetuated this linear, moralistic worldview—a view sponsored by those who stood to gain

41

most from maintaining public faith in traditional values and allegiances. By the 1960s, however, a few events conspired to change the tube's role from an enforcer of linearity to a promoter of chaos. The Kennedy assassination, for one, cannot be underestimated for its long-term effect on the media public. Culturally, of course, the assassination marked a lapse in the continuity of our government. Yes, according to protocol Lyndon Johnson assumed his rightful place in the Oval Office, but a term we had collectively set in motion with our votes had been abruptly ended. More insidiously corrosive to our sense of continuity was the way our media deconstructed the moment-to-moment reality of the assassination and its aftermath.

The Zapruder film, a watershed event in American television paralleled only by the Rodney King tape and maybe O. J. Simpson's Bronco ride, took an already discontinuous event and broke it down further. We watched, frame by frame, as our president's head flung back, and his wife sprang out of her seat onto the trunk of the limo. Rather than providing us with answers, our news media unintentionally flooded us with questions. Was Jackie escaping the car? Was she reaching for a piece of her husband's brain? Was there only one gunman? Two? A conspiracy? Making matters worse, when the suspected gunman was finally being escorted to his formal and customary arraignment, he was murdered, too! We would get no satisfactory ending. No matter how much tension was generated, relief was nowhere in sight. The discontinuous style in which the information was presented to us, coupled with the overwhelming discontinuity of the event itself, altered our relationship to the image on the set and our image of reality.

There were three cultural reactions to this tear in linear reality. Some chose to ignore it completely. These people convinced themselves that even though the Kennedy assassination appeared utterly discontinuous, it was actually a simple and linear event. The assassin got killed a little early, Jackie panicked a bit, the camera footage was blurry, and FBI physicists figured out how a bullet can zigzag back and forth. In the long term this was, perhaps, the most painful path to take. The resignation of Nixon, the end of the cold war, and the development of a postmodern culture has left these people in a state of unimaginably irreconcilable cognitive dissonance. The second reaction was to try to create connections between discontinuous events, even when there were none. These were the conspiracy theorists (we'll be seeing a lot of them in later chapters) who were compelled to create an antiauthoritarian but nonetheless linear explanation for what had happened to their president, nation, and values.

The conspiracy theorists were the baby-boomers, who brought us some tremendously positive social transformation. But their problem was that they still believed in the power of authority and value of static moral templates. They staged antiwar protests, civil rights demonstrations, and hippy-yippy revolutions because they saw the incumbent regime's templates as wrong and their own as right. They regarded the president and their teachers as powerful parent figures who needed to be revolted against. Their peculiar style of revolution was to revise the way that lines of continuity were being drawn among inherently discontinuous events. It's like a game of connect the dots. The young baby-boomers wanted to change the numbers on the dots, so that the lines would be drawn in a different order and the picture would come out different. When this form of conspiracy theory works, it yields Woodward and Bernstein's Watergate investigation. When it seeps out to popular culture, it yields an Oliver Stone movie. All this, again, in reaction to the first incursion of media discontinuity into mainstream social awareness.

But there was a third reaction to the Kennedy crisis—the reaction of children growing up with this assassination as their first presidential memory. To anyone under thirty-five,

43

presidents are, by definition, people who get assassinated. To them, the Zapruder film is a media classic. This generation does not turn to the media for answers, but for questions. They understand that to subject a seemingly linear event to media scrutiny is to break that event down into its component, frame-by-frame segments. It loses its original meaning— or at least the meaning that may have originally been intended. For this younger audience, discontinuous media is not the exception, it is the rule. As a result, they have adopted a social philosophy very different from their predecessors'. They do not work to recombine and reduce the seemingly endless stream of media bits into coherent, unified pictures, and they no longer believe in hard-and-fast answers to the world's problems.

They are comfortable in the disassembled mediascape because they are armed; the first weapon to appear in their arsenal was the remote control. As the media became increasingly chaotic, the remote emerged as the surfboard on which the armchair media analyst could come to reckon with any future attempts to program him back into linearity. The minute he felt the hypnotic pull of story or propaganda, he could impose discontinuity onto the flow by changing the channel. At first, this served as a simple negation of the broadcast reality. Whenever a newscaster says something irritating to the viewer's sensibility, even an Archie Bunker can curse at the screen and change the channel in disgust. Eventually, though, in order to maintain their viewerships, the media dispensed with linearity altogether, and devised new styles and formats to appeal to the channel surfer. They didn't realize what they were doing.

The media's own form of discontinuity, the edit points linking one shot to another, have been around since the films of D. W. Griffith. When film editing was first introduced, moviehouse audiences were baffled by a visual language they did not yet understand. When a film would show a house

44

burning down, then cut to a reaction shot of the homeowner standing outside, the audience had no idea what they were supposed to be looking at. Every edit point was experienced as a break in continuity—a lie. The art of moviemaking was learning how to create stories so compelling and geographies so ordered that the audience could leap the gaps created by each cut. By slowly developing a series of conventions like size of shot, axis, eye-lines, and frame composition, the moviemakers created a language called "film grammar." Anyone who can watch a film and understand what's going on has learned this language of the edited moving image.

Over time and, eventually, to make television more compelling to the remote control viewer, television editors sliced and diced their programming into more rapid-fire segments. Although any film textbook will explain that an audience cannot comprehend a shot with a duration of less than two seconds, and that such rapid-fire editing should be used only for effect, two-second shots and even one-second reactions soon became the norm on television. Like the drips of water coming out of a faucet at an increasing rate, once the speed of edits reached a critical frequency, the linear story just broke apart as the programs reached turbulence. The media chaos this turbulence generated was called MTV.

Music television is a celebration of the gaps. The component segments of a rock video fly by too quickly to be comprehended on an individual basis. MTV must be thrashed as if on a skateboard. The texture of the programming is more important than the content. The rapidity of edits produces a new sort of changing image. Just as a regular film is made up of thousands of frames running by so quickly that it creates the illusion of a single moving image, MTV juxtaposes its moving images so quickly and so disjointedly that it creates another level of imagery. This style of rough, disjointed media was precisely the landscape preferred by the channel surfer. It made coercion through traditional, narrative programming tech-

45

niques impossible, and required that a new language—a language of chaos—be developed. The kids watching MTV learned to speak it like natives.

Advertisers had the greatest stake in adopting this language quickly into their repertoire. From watching MTV videos, they learned to make commercials with rapid cuts, screens within screens, and, most important, no linear story. Their products couldn't look like conclusions to a set of propositions but instead needed to nestle themselves within the gaps of an intentionally disjointed set of words and images. Their rapid and disjointed editing style was soon adopted by primetime programmers as well, who developed shows like *Hard Copy*, *COPS*, and eventually *ER* and *NYPD Blue* (their very titles evoke discontinuity). Now the viewer with his remote was no longer the only one imposing nonlinearity onto television programming. Like a self-deprecating Borscht Belt comedian, the shows were doing it to themselves.

It can be argued that these postmodern qualities of television are nothing new. True, cubists, dadaists, situationists, and others have long understood the advantages of cut-and-paste techniques and employed them in their work. Writers from James Joyce to William Burroughs used cut-up language and juxtaposed samples in order to create confusion, resonance, and persuasion in their texts. But never before has such dislocated imagery been in the mainstream. These formerly ingeniously unconventional methods of expression are now absolutely conventional.

A few other innovations conspired, so to speak, to make the mediaspace into a full-fledged dynamical system. Cable television, for one, increased the number of channels in the home from about six to over fifty. The number of gaps to be surfed increased accordingly, transforming the mediaspace from a set of streams into an ocean break worthy of championship tube runs. Even more significant, home media like

46

camcorders, video decks, and, later, satellite dishes, public access channels, computers, modems, and networks changed the essential rules of the mediaspace. Media was no longer something to be passively received. It could be created and, to some extent, broadcast by anybody.

Many media analysts, including myself, have outlined the way our media evolved from a top-down, unidirectional forum into the interactive free-for-all it is today. Some explain how the media used to be "one-to-many," meaning that just a few broadcasters controlled what millions of viewers absorbed, and go on to rejoice that now the media is "many-to-many," because anyone with a modem or camcorder can tell his own stories to the rest of the world through public access television or the Internet. Anyone can feed back to his leaders, thanks to call-in radio or forum-style talk shows and presidential debates. George Bush lost the 1992 election because Clinton could respond better to the demands of the interactive media marketplace.

While all this is true, it is still an oversimplified view of the way our media has reached its current level of complexity. There has always been feedback in politics and government. It has just operated a lot slower than it does today. If a king or president enacted too many policies against the public interest, the people would revolt or elect someone else. Even if the public had been bamboozled by the public relations strategies of the day, eventually, if taxes were too high or war too tragic, they would change things. Consumers' demand still affected the producers' supplies, and governments had to keep their populations contented enough so that they would not rebel. Just the fact that leaders and marketers employed public relations strategies proves that they understood the natural laws of society, however suppressed or delayed the reactions. The introduction of home technology into the mediaspace increased the flow of feedback from the "many" back to the "few." Once enough interactive channels had been put in place, that flow, like the content of television media, reached turbulence.

The watershed event that made us all aware of the media's new turbulence was the Rodney King tape. Even without an Internet to speed our opinions around the globe, one man with a $300 camcorder captured a moment of injustice on tape, and the chaotic nature of the mediaspace spread that tape, like a virus, throughout our culture overnight.

Were the media not already exhibiting the properties of a dynamical system, the Rodney King tape would never have replicated as it did, nor would it have provoked such a violent social upheaval. The format of television had reached turbulent levels and consisted of little more than discontinuous bites of data and imagery. The Rodney King tape could be effortlessly inserted into every newscast, tabloid show, rock video, and even the opening credits of Spike Lee's *Malcolm X*.

Had media technology not reached turbulence, there would not, in all probability, have been a camcorder-ready individual within earshot and camera range of the scene of the beating, nor would there have been such an instantaneously responsive mechanism for the tape's distribution. A television producer, somewhere, may have even decided against airing the potentially incriminating evidence.

Had the media not developed into a natural system, capable of responding to our societal needs, the tape would not have caused even a stir. Rodney King would have meant very little if blacks in America weren't already subject to brutality of many kinds. There is a police problem in this country, and there is a race problem in this country. They just aren't being addressed appropriately yet. A media virus like the Rodney King tape will only spread if there is a cultural immune deficiency to the viral code within it. In this way, then, like a true partner in our cultural evolution, the chaotic media serves to promote or at least bring to the forefront the agendas that we need to address.

Finally, though, if we as members of the chaotic mediaspace were not equipped to absorb the data flying across our screens, make sense of the postlinear grammar with which it is formatted, and participate in its production as amateur journalists, then our media's ability to promote our cultural evolution would remain totally unrealized.

If we couldn't surf the waves of modern media, we would surely drown in them.

But the skills we need to develop in order to become adept at surfing channels (and computer networks, online services, the World Wide Web, even our own e-mail messages, for that matter) are the very opposite of what we traditionally valued in a good television viewer. We are coming to understand that what we so valued as an attention span is something entirely different from what we thought. As practiced, an attention span is not a power of concentration or self-discipline in the least, but rather a measure of a viewer's susceptibility to the hypnotic effects of linear programming. The "well-behaved" viewer, who listens quietly, never talks back to the screen, and never changes channels, is learning *what* to think and losing his own grasp on *how* to think. This was a gap-evasive viewing style that ignored the basic reality of a discontinuous mediaspace, helping us convince ourselves that our lives could run smoothly and easily if we simply followed instructions.

We see now, though, that the viewing style of our children is actually the more adult. Nearly every essay about kids and television cites the (relatively undocumented) fact that the attention span of our children is decreasing dangerously. In a brochure from the famously progressive Rudolph Steiner/Montessori schools, parents are encouraged to prevent their children from watching any television at all. Even *Sesame Street*, this Luddite document contends, is formatted with so many quick edits that it decreases the attention span. On the contrary, the ability to piece together meaning from a discontinuous set of images is the act of a higher intellect, not a lower one. Moreover, the child with the ability to pull himself out of a linear argument while it is in progress, reevaluate its content and relevance, and then either recommit or move on, is a child with the ability to surf the modern mediaspace. He is also immune to many of the methods of programming and persuasion foisted so easily upon our unsuspecting, well-behaved viewers. He refuses to be drawn in as a passive, receive-only audience member.

The child of the remote control may indeed have a "shorter" attention span as defined by the behavioral psychologists of our prechaotic culture's academic institutions (which are themselves dedicated to little more than preserving their own historical stature). But this same child also has a much *broader* attention *range*. The skill to be valued in the twenty-first century is not length of attention span but the ability to multitask—to do many things at once, well. Remote control kids can keep track of ten or more programs at once, and they instinctually switch from channel to channel just in time to catch important events on each one. If several important shows are on at the same time, the remote child does not need to tape one for later viewing. He watches both at the same time, because during that "later," there will be something else to watch or do.

The other key viewing skill that kids have developed, which may be linked to the so-called shortened attention span, is the ability to process visual information very rapidly. A television image that takes an adult ten seconds to absorb might be processed by a child in a second. Certain MTV videos and Saturday morning cartoons are utterly incomprehensible to adults, who must, in a step-by-step fashion, translate each image as well as its subcomponents into a language they can understand. Kids process the visual language of television directly and are so adept at this skill that they hunger for images of greater complexity. The television imagery of MTV and kids' commercials consists of frames within frames, text mixed with pictures, multiple images, and juxtapositions of images at the freqency of a strobe.

This is a new language of visual information, and it depends as much on the relationship of different images and images within images as it does on what we generally understand as content. It suspends the time constraints of linear reasoning in order to allow for a rapid dissemination of ideas and data, as well as the more active participation of the viewer to piece it together and draw conclusions for himself. If anything, this development would indicate an evolutionary leap in the ability of an attention span to maintain itself over long gaps of discontinuity, either between channel surfing cycles, or from session to session.

In the workplace of the future, a broader attention range and shorter absorption time will be valuable assets. The stockbroker with a broad attention range will be able to keep track of many markets at once as they flash by on his terminal. He will be able to talk on the phone to a client on one line, his boss on another, and an electronic "chat" on his computer screen, all simultaneously. His shortened attention span will keep him from getting too unconsciously engrossed in any one conversation or activity, and always ready for something new.

Meanwhile, the ability to process visual information quickly will enable the worker of the future to cope with the "information overload" our data highway nay-sayers are busy warning us about. If we are about to enter an age of information glut, those who can wade through it will be people with the ability to inspect, evaluate, and discard a screen of data immediately. This information skimming will need to be practiced on many different levels, and sometimes simultaneously. A person might scan his online services for the one that has a high volume of new messages. Once logged on, he will scan his list of incoming messages for the ones that might be important. Once inside a message—say, an electronic newsletter—he will scan within it for articles, then paragraphs of relevance. He will then need to file or forward it to the appropriate place for later use, if any.

Like a surfer who comes to understand the self-similar quality of beaches, tides, waves, and their components, the data surfer comes to recognize the qualities of different sorts of data structures. A television image directed at a sports fan or automobile purchaser will have a great deal of horizontal motion. (In the sports event this results from the camera moving back and forth across the screen; in the automobile ad it comes from panning with the car across the highway.) The quality of the image—which can be gleaned almost instantaneously—is enough for the channel surfer to decide whether to hold still. An experienced surfer keeping track of sports scores may only stop on the sports station when a graphic image of the score is onscreen. As he skims by this channel, he watches the bottom of the screen, because he knows this is where that graphic material will be superimposed. The same sort of multitiered scanning will be an essential skill in wading through electronic mail. The "subject line" of an e-mail message in a list will have certain elements if the message is important. The important paragraphs within that message will have similar qualities, drawing the experienced recipient's attention. Like Beavis and Butt-head, the executive of the future will be able to determine which e-mail is "cool" and which "sucks" in a fraction of a second.

This ability to recognize the quality of something from its shape and to trust one's impulses based on this recognition, may be the key skill in understanding any chaotic landscape. We all find nature so reassuring because we can recognize the self-similarity of its systems. The leaves of a tree have structures resonant with the branches, which are resonant with the other trees. Any truly chaotic system has such patterns, which, once recognized, provide a way to understand the system on an intuitive level. To recognize the pattern is the first step. To take the leap of faith that this pattern is correct or even that it exists, is the second and more dangerous step. But, as the snowboarder advised, it is usually always better to overshoot than to underestimate.

Understanding the modern language of visual information and coping with the culture it has spawned is an ongoing process. I was once in a crowded bar watching a basketball game on TV with the sound turned off. During a time out, most of us began drinking our beer and conversing with each other. All of a sudden, an older woman jumped up on her stool and screamed "Yes!" We all turned to her, confused, until we realized that she was watching a replay of a rather spectacular moment from last night's game. She sat back down, embarrassed. Because she did not understand the grammar of video replay in a basketball game (a small frame expands to fill the screen before the segment is shown) she had no idea that she was not watching "live" television.

For most people, sports bar skills are not of paramount importance in achieving life success, but the inability to process and interpret visual media will soon be much more debilitating than many of us care to admit. Rather than locking up our television sets and crippling our children's ability to compete or even participate in the mediated culture ahead, it is we who should take a remedial course in postlinear, visual communication.

We should read comics.

Playing in the Gutter

Kids who read comics have a better head start understanding media than a student of McLuhan himself. Comic books may seem visually oversimplified and thematically primitive to most adults, but these very qualities are what permit an active participation from their readers and the purposeful manipulation of time and space by their author-artists.

Comics vary widely in their pictorial complexity, from line drawings of stick figures with no backgrounds to realistic, detailed paintings of characters in panoramic settings. Generally, though, however elaborate their style, comics depend on iconic representation rather than illustrative depiction. That is, they communicate with basic symbols and relationships, and the rest is filled in by the audience. Charlie Brown's head, for example, is little more than a smiley face, but we as an audience know his psychology so well that we can fill in the rest of his facial expressions by ourselves. By omitting everything that has no direct bearing on their stories, comic book artists can more easily direct our attention to particular changes in plot, emotions, or location.

For clarity, and against adult intuition, the parts of a drawing with the most impact on the plot or characters tend to be the most simply drawn. The starkness of the image makes it stand out, while its iconic quality makes its meaning easy to recognize. In old Warner and Disney cartoons, backgrounds were painted in detailed watercolors, while Bugs or Mickey was overlaid as a simplified "cell" drawing. This technique arose mostly out of financial considerations—it was less expensive to paint one elaborate, static background than to repeatedly paint frame-by-frame detailed portraits of characters as they moved. When this style crossed over into comic books, it amounted to a leap in their ability to communicate.

Icons transmit important information better and more quickly than detailed pictures. Think of an international road signal for "Don't Walk." It consists of a hand, palm toward us, in the position of a policeman's hand telling us to wait. The picture doesn't really look like a hand, nor do we know whose hand it is. It is a generic hand. Because we recognize the icon and that the color, red, generally means "STOP," we understand the icon. If it were contained within an octagonal frame, better yet for those of us who associate that shape with a stop sign. If it were a full-color scene of a policeman holding up his hand and people stopping, the sign would be much harder to recognize or translate. Those details are irrelevant. An icon condenses information so it can be scanned and understood quickly.

In this sense, icons amount to the recognizable patterns underlying more complex situations. They depend no more on an artist's ability to isolate them than they do on the viewer's ability to recognize their meaning and infer their greater context. The simpler an icon, the more universal its application. If an icon is too simple, its meaning can become ambiguous, but if its rendering is too detailed, it becomes too specific and less widely applicable. The first time we encounter an icon, we need to interpret it and its context. A sign showing a fork and knife could mean a cutlery store, but if it's in an airport, chances are it indicates a place to eat. Once we are familiar with the icon, we only need to determine the way it applies to the situation at hand—the icon in its context.

(Interestingly, this is analogous to the conceptual breakthrough that has made video teleconferencing possible. Programmers needed to find a way to compress moving video images to reduce their bandwidth to manageable proportions. They accomplished this by isolating the parts of the picture that move from the parts that don't. Modern teleconferencing technologies transmit only the pixels of the image that have moved, like the speaker's mouth or eyes, and update just those portions of the video picture.)

Facility with iconographic representations depends on our iconic vocabulary and even more on our ability to recognize the primal, elemental structures within our fully detailed reality. Just like the ability of a surfer to recognize certain wave patterns, or the skill of a mathematician to recognize a repetitive, fundamental shape within a fractal, the ease with which we can recognize the basic elements in the language of visual gestures determines our ability to navigate this nonverbal, nonlinear landscape. This requires both an experienced eye to isolate a pattern and a flexible mind to generalize that pattern to a different level of awareness to apply it to real life.

Comic books communicate almost exclusively through icons, both educating readers in a language of visual patterns and depending on them to recognize new patterns or new combinations of established ones.

The iconizing of characters in comic books shapes the way we relate to them. Artist Scott McCloud proposes in his terrific comic book, *Understanding Comics*, that the simplicity of comic book characters is what allows children to identify with them so completely; this also explains the success of comics across cultural and national borders. The simplified face of the character is more universal, so a child can imagine himself as that character, filling in his own qualities to complete the picture with his imagination. This is why animals are such effective cartoon characters—their human qualities can only be represented iconically, since mice and rabbits don't show facial expressions in real life. If a character is drawn in great detail, McCloud explains, he becomes a more specific, defined human being. Instead of entering the character's world and seeing it through his eyes, we look *at* that world. Our attention is also drawn away from the actions as we are forced to regard more than one level of detail.

In the world of comic books, who a character is usually matters less than what he does and says. The only exception to this are villains, who tend to be drawn in greater detail so that we know more about them from the outset. We then look at the villain, as our protagonist does, in a more objectified fashion. Note this is done without the benefit of a lengthy, time-based back story detailing the villain's crimes, but rather with a few immediate details and an overall stylistic cue. Consider the difference between the simple way Disney's Snow White is drawn compared with her enemy, the evil queen, and you get the idea. The depth of the evil queen's "history" is represented in the length of time it takes to absorb her image. We identify with the simply drawn Snow White, and, along with her, observe her more complex antagonist from an objective distance.

In the language of comics, actions, emotions, and events are also represented iconically. Motion is indicated by a few curved lines. Do objects in the real world have lines trailing them when they move? No, but we have come to accept the convention for its expedience, as well as the freedom it grants our imagination to fill in the quality of that motion. A character drawn small in the bottom corner of a frame looks alone and dejected, and one whose face fills the entire panel is understood as experiencing an emotion. A primitive way to iconize an idea is with a light bulb over the character's head. Icons tell the story.

This is one of the ways iconic representations free the comic medium from the constraints of linear storytelling and thus train comic book readers to understand the world in new ways. Unlike film or television, which usually try to smooth out the gaps between edit points, comic books exploit those gaps in order to communicate. The sequence of panels in a comic book and their internal organization communicate as much as the individual icons within them. Take the example above, where a character gets an idea. A more sophisticated way to communicate this event would be through a sequence of two or more panels, contrasting the character's facial expression or body posture before and after coming up with the idea. A comic cannot show a character raising his eyebrows or widening his eyes the way a movie can. Comics don't move. A comic book must communicate through a disjointed series of images. The images are static and separate; it is up to the reader to understand them as a coherent whole. But again, unlike a movie or other linear format, this is not done by smoothing out the space between the images, but by highlighting their differences. Before the idea; after the idea. The moment that the character comes up with the idea occurs in the space between the two frames, the empty space called the "gutter." The meaning lies in the gutters, when time actually passes.

Because of icons, a single frame can also contain internal gaps that communicate the passage of time. If a comic frame were, like a traditional photo, a frozen moment in time, then it would be impossible to have two characters speak to each other in the same frame. But, as we have all seen, two dialogue balloons can appear within one picture without causing us too much confusion. As long as the artist follows the convention of laying out the action from left to right (in Western comics, anyway) our eye intuits the order in which the people are speaking.

A comic frame can condense much more time than this. Let's say our character who comes up with an idea is a scientist—Archimedes discovering the famous law of hydrostatics. A single page-size frame might show a picture of the rooms of Archimedes's whole house, with Archimedes in each one, working on the problem. One Archimedes is coming in the door, intent on solving the problem assigned to him. Another is working in the study. A third is in the kitchen experimenting with a glass of water, and so on. The last Archimedes is in the bathroom, having given up on the assignment, just stepping into the tub. Turn the page and he's flying out his front door shouting "Eureka!" All the different Archimedeses were not existing simultaneously in the same house, but by putting them all together, the artist can communicate a lot about the scientist's state of mind. No matter where he goes, he still has no answer. He's still locked in the same frame, which, although it covers many different times, only contains one main psychological state.

A single frame can contain the pitch of a baseball, its delivery, the batter hitting it, and the ball exiting the park. Time and space are malleable, and most of the meaning comes through the ways these rules are broken, and contradictory realities are juxtaposed. Even the text within the frame contradicts the temporal reality of the drawing, but we learn to accept a round speech balloon, a cloudlike thought balloon, or a rectangular narration box as conventions. Comic books do not demand that we dispense with linear reality altogether. There is still a story to follow, a hero with whom to identify, and a climax to anticipate. But we must do all this by instinctually recognizing and interpreting a series of iconic representations while analyzing their discontinuous arrangement and greater context. This is the task of finding meaning and continuity within chaos.

The comic book industry is itself like a dynamical system. Because comics can be created with a low-cost, low-tech skill—drawing—yet distributed widely by a massive publishing system, the comics industry provides great amplification for the visions of individuals. Any kid with good drawing skills and an interesting idea can exploit his high leverage point by attending a comics convention with samples of his drawings and getting a contract with a major publisher. Others use photocopiers to get their work out by any means necessary, and sometimes hit it big. The comics industry is delightfully unpredictable because, in a business, chaos means opportunity.

Like the language, format, and business of comics, its fictional worlds are accordingly chaotic in nature. Perhaps for business reasons at first, the editors at Marvel and DC comics invented the convention of crossover plots. This way, Superman might chase a villain out of his own comic book world, but the villain could reappear in Spiderman's. The comic book editors get to reuse a character, continue his own story without introducing him all over again, and, if all goes well, expand the readership of each comic book. Readers of Superman will know that certain plotlines have expanded beyond the boundaries of their own book and feel compelled to follow them. If they get hooked on Spiderman, they will follow him, too, when he makes a guest appearance in Thor, and so on. This strata of cross-referenced superhero worlds was expanded and more fully conceived by Marvel's Jack Kirby and dubbed the Marvel Universe.

Kids raised in the Marvel Universe may as well have been born in a fractal stew. One of the more bizarre aspects of this universe is that the superheroes appearing and working together are often from different historical time periods or levels of fantasy. One might be a policeman, another a dragon, and yet another a robot from the future. Even the styles in which they are drawn may differ dramatically. For these characters to meet in one comic space, writers must invent interdimensional disasters that break the rules of historical time and space, demanding the attention of superheroes from many different planets, eras, and cultures. Although they are divorced of their original contexts, the characters maintain their own iconic identities and traits. The joy of these special issues, which usually become valuable collectors' items, is to watch the different superhero skills working in tandem. Now Thor could team up with the Fantastic Four to defeat a common enemy. Just as there can be many icons and moments within a single frame, so can there be many superhero worlds within a single comic book.

To accept these conventions, though, the children reading comics dispense with the rules of linear order and continuous logic for a much more rewarding sequence of icons, gaps, and discontinuous relationships. By abandoning the immediate soothing smoothness of a linearly consistent world, they achieve a greater continuity—a metacontinuity, if you will— on another level. Only a child who has a true, long-term attention span will remember that when The Hulk met an adversary in, say, issue 123, he mutilated the villain's left arm. When the same villain shows up in *The Avengers* with an apparently healthy arm but wearing a strange black glove, the arm hidden beneath it must be a bionic replacement. But when he shows up in *Thor* with two good arms, they are both the originals, since *Thor* takes place centuries before his injury.

To understand continuity in the way our children do, we must rescale the way we search for it. The apparent islands of discontinuity really are, like an archipelago, joined to one another beneath the surface of the water. And the only way to know that for sure is to go under and see for ourselves.

2

THE FALL OF DUALITY AND THE RISE OF HOLISM

"Getting old is a terrible thing. I remember when I realized I could beat my dad at most things. Bart could beat me at most things when he was four."

Homer Simpson, on repeatedly losing
to his son, Bart, at video games

Spending time in the Marvel Universe changes the way you think about the world, probably for the better. While comic books are best known for their broad-stroked plots and superheroic characters, the particularly nonlinear experience of comics conveys a complex, richly textured, and internally consistent world.

Marvel Universe creator Jack Kirby helped the comic book pantheon evolve from a series of separate threads into a web of interdependencies. Likewise, his own comic book universes evolved both structurally and thematically to incorporate increasingly complex and subtle elements. What began as confederations of superheroes allied against their common enemies eventually developed into a tapestry of mythic forces and entities, whose struggles and identities proved much less clear-cut. Instead of just watching handsome Superman conquer his gruesome and evil monthly enemy, we learned to travel with Swamp Thing, himself a rather hideous monster, through the roots and vine of the earth's own plant life as he attempted to reckon with distortions in the balance of nature and his own rage at the human perpetrators of ecological decay. The comic book universe became almost disorientingly self-similar, with the structure and development of the comic book world reflected in the structure and development of its plots and themes.

The development of artistic and narrative form in media and stories for children reflects an overarching evolution in themes and cultural values. This is most obviously expressed in the way the superhero persona changed over time. The most popular comics of the World War II era, like Superman, reflected very simple themes for a nation whose chief problem was not deciding how to judge evil, but how to conquer it. Superman's red-and-blue palette stood for American patriotism, and even though he was both an immigrant (from planet Krypton) and a member of the media (*The Daily Planet*), he could be counted on to defend and protect our flag as if it were his own. Interestingly, his one weakness—the susceptibility to an element called Kryptonite found on his home planet—reveals a lot about our greatest perceived risk in the early half of this century: that immigrants will show allegiance to their native countries rather than their new host. The "American way" Superman fought for was dependent on people denying who they really were in order to dissolve into the great melting pot. Superman hid his own true ethnic identity and worked to stymie the *Daily Planet*'s efforts to get a scoop if it would endanger anyone or anything he deemed best kept from the public's view.

The postwar generation got Batman. His stories were still intensely linear but plagued with self-doubt. Batman was a baby-boomer, struggling to do the right thing. Guilt-ridden from the get-go, he battled with the inconsistencies of his supposedly dualistic world. He watched his parents get murdered by muggers but inherited millions of dollars from them. He experienced only the worst extremes of his increasingly bipolar reality, donning a smoking jacket for his daily upper-crust soirees and his bat cape for nightly expeditions to the mean city streets. Meanwhile, like the limousine liberal passing through impoverished slums on his way home from work, he struggled with the building cognitive dissonance between his own wealth and guilt over its origins. Desperately serious, Batman is easy pickings for the Joker or the Riddler, who can always generate fun at Batman's expense. Batman returned in the cyberpunk era as the Dark Knight, a near-crazed crime fighter who is being driven insane by the inability of his dualistic moral template to interpret the postmodern swirl around him. TV commentators in frames within frames intrude on every page, questioning Batman's actions, motivations, and even his sanity. As a dualistic and self-doubting boomer, Batman is unprepared for the modern social landscape.

The heroes of the biggest screenager comic book sensations are themselves kids of chaos: Teenage Mutant Ninja Turtles and Tank Girl, who have dispensed with dualistic judgment altogether and trust their instincts to know whom to beat up and how. The object of the game is to have as good a time as possible, eat pizza, make explosions, and surf one's way (cowabunga!) through life. In these comics, it is the villains who are so desperately serious, and the good guys joke at their expense. These comics celebrate what Batman lamented and Superman repressed: individuality, weirdness, inconsistency, openness, and even mutation.

Superman and Batman were secretive about their identities. They were split personalities, paralyzed by the fear of their own discovery. What is it that they would really have surrendered by admitting their identities? Secrecy is the way to maintain duality. It is what we find at the core of the separation between scientific kabals and religious kabals, the mind and the body, and, finally, the paranoid clinging to a linear interpretation of our world. Secrets maintain the separations and categories—the ignorant and knowledgeable, the victims and masters.

Traditional linear stories tend to express themselves in duality. A cause leads to an effect. A hero fights against a villain. The forces of good attempt to extinguish the forces of evil. Good either triumphs or fails. Resolution is the object. Like the end of any *Scooby Doo* episode, we pull the mask off the monster to discover his true and distinct identity. End of story.

Nonlinear stories tend to express themselves differently. Because they don't define causes and effects, heroes and villains, or good and evil, they cannot conclude in clear-cut triumphs or defeats. The force driving the story is no longer a quest for an objectified resolution, but an ongoing evolutionary pressure. Like the famous multiple endings of a *Wayne's World* movie, the *Scooby Doo* scenario is only one of many possibilities. The story just goes on.

The force that pushed the world of comics from linear simplicity into nonlinear complexity was the same as for any chaotic system: turbulence. As single stories became connected to one another through crossover plots, shared characters, locations, and themes, the linearity of their plotlines was sacrificed to interconnectivity. As more and more connections formed, and characters could be appreciated in relation to one another, it became impossible to maintain the polarized extremes of black-and-white conflict. As the comics universe took on the qualities of a chaotic system, a new, self-similarly antipolarizing theme emerged: Dualities are false, and evolution is the driving force of nature.

Kirby's own crowning achievement, the miniuniverse of his book *The Eternals* (1970s), directly addressed these issues. According to the back story, the Eternals are the fully evolved humans responsible for planting DNA into precivilized man. Now they have returned to judge mankind's evolutionary progress. If we succeed in reaching the next phase of evolution, we will join the Eternals as cosmic beings. Working against the efforts are a race of primitive humans known as Deviants, whose unstable DNA has led to their rather uncivilized practice of killing any of their evolutionarily mutated progeny. The Eternals, on the other hand, remain linked together through a kind of collective consciousness called the Uni-Mind: "Human conjecture CANNOT grapple with what is happening, the concept of the Uni-Mind is unheard of. . . . Yet . . . it exists among the Eternals!—and is emerging from its fiery spawning ground. In full view it is like no other life form on earth. It is a living group organism! The only [one] of its kind. . . . The Uni-Mind lives in the climate of catastrophe."* If we can progress enough in the next few decades, we, too, can join the Eternals as part of this great group being.

In the kids' world of comic books, the notion of apocalypse—a result of overpolarized ideological extremes—is replaced by the concept and dramatic experience of an evolution-based colonial organism. There is a way out: embracing chaos, and changing as a result.

New Types

Maybe none of the world's cultures has confronted the reality of apocalypse more directly than World War II Japan. And as we would expect, their comic book and animation traditions were called upon to answer some pretty pressing questions about the end of the world as we know it.

* Jack Kirby, *The Eternal*, no. 8 (Marvel Comics).

As experienced by the Japanese, however, the main polarity to be reconciled in order to avert doomsday was between nature and technology: man and his dangerous inventions. Years before environmentalists warned the West about the ecological repercussions of our technological strides, the Japanese had endured a nuclear holocaust. While Jack Kirby's cohorts were musing on the nature of superheroism as it relates to postmodernity and the discovery of DNA, the Japanese were digging themselves out of the ruins of Hiroshima. Forced to consider the real possibility of technology destroying nature altogether, they struggled with the implications of this duality on an accelerated learning curve.

With only one movie studio still in operation after the war, the Japanese turned to *manga*—serialized comic books—for visual entertainment and, more important, for pop cultural answers to societal dilemmas. The lengthy, almost Dickensian stories of manga often took years to reach their conclusions, if they did at all, and stretched their readers' capacity to tolerate continuity on such a grand scale. By the time the studios arose to turn these comics into animated features and television programs, the Japanese audience was ready to accept epic plots spanning several films or entire television seasons. Part of the thrill was to watch characters and plots evolve along with the style of animation.

The Japanese studio system promoted this evolutionary approach. Younger animators earned internships with established artists and learned to replicate the original artist's stories and drawings before slowly introducing their own visions. The studio system also provided a safe haven for many artists with extremely unconventional lifestyles and points of view.

Osama Tezuka, Japan's equivalent of Jack Kirby, is considered the father of the Japanese animation style kids still watch today, animé. A big Disney fan, Tezuka drew characters with large eyes and oversimplified features. But his television series dealt with issues and characters that wouldn't be mistaken for the Magic Kingdom's. His first big success, *Astro Boy* ("Mighty Atom" in Japanese), featured a cute, tiny atomic-age hero—a boy who gave Japanese children of the 1950s the chance to identify with someone just like them, except for his ability to tame atomic energy. The 1950s also brought *Gigantor*, featuring a giant robot who fought monsters and abusers of technology and was directed by a tiny animé boy with a remote control. The young animé hero of *Speed Racer* was the only driver capable of exploiting the many space-age features of the Mach Five in order to foil the efforts of mean adults with other cars or technologies designed for crime.

Like Tezuka's animé characters, who were kids capable of harnessing the sometimes terrifying applications of technology, the children of the famous rubber-suited monster movies, too, were the only people capable of understanding the inner struggles and gentle natures of Godzilla, Mothra, and kiddie favorite Gamara. The monsters themselves were mostly the mutant result of nuclear accidents or toxic spills. The kids, because of their innocence, could communicate empathically with the monsters, sing to them, and to some extent control their actions. Kids could even persuade monsters to help save Tokyo from alien invasions or other disasters. Adults accidentally create monsters and catastrophes by letting their technology get out of control, while the children—thanks to their ability to understand the secret workings of technology and the secret hearts of monsters—are uniquely qualified to clean up the mess.

By the 1960s and early 1970s, these themes evolved further along with the audience's tolerance for increasingly complex plot structures and levels of continuity. As the plotlines of series extended over multiple seasons and among different series, animé shows began to reflect some of the properties of dynamical systems and to apply these principles to their preoccupation with evolution, technology, and children.

In the mid-1970s, while George Lucas was developing *Star*

Wars in the United States, animé's Leiji Matsumoto was hard at work on his own epic space saga, beginning with *Arcadia of My Youth* (titled *Space Pirate Harlock* in Japan). He continued and expanded on the themes of his predecessors, but confronted the World War II disaster much more directly. In Matsumoto's mythos, the earth has been destroyed by aliens. Space Pirate Harlock, a reincarnation of a samurai warrior and a World War II Nazi pilot, as well as a recycling of one of Matsumoto's favorite manga creations, fights the enemy aliens. His comrade, another of the artist's rep company of characters, knew Harlock in his previous incarnations as samurai and pilot and once again serves as the loyal vassal and aircraft engineer. The theme of reincarnation, both of the samurai warriorship and WWII warriorship, hearkens back to ages of Japan's former greatness. The fact that characters return and are recycled time and time again instills the notion that something essential in the personality of the Japanese people, or earthlings themselves, can never be completely destroyed, and will certainly recur.

The recurrence of shapes and patterns over time is another of the more spectacular qualities of dynamical systems. Analysis of genetics, weather patterns, and populations demonstrates that no matter how wildly a system swings from extreme to extreme, some measure of its original character is retained and can pop up, in an identical form to its original, years or centuries later. It works almost like a strobe light or a movie camera. Did you ever see the hubcaps of a moving car in a movie or TV show? When the car first moves, the design of the hubcap blurs, but as the car reaches a reasonable speed, the design on the hubcaps reappears, as if they are not rotating at all. This is because of the flashing frames of the film that only capture the image of the hubcap in certain positions as the wheel spins. When a system reaches turbulence—and this applies to cultures as much as hubcaps—qualities that were thought to be lost reappear with striking clarity. This is how a turbulent mediaspace fosters the recurrence, or "recursion," of styles and attitudes from earlier decades or centuries. Heavy metal brought us medieval poetry, and Industrial music reiterated Goth.

Well, the repetitive complexity of the Japanese animé industry produced just such turbulence and recursion. Matsumoto kept reusing his characters—not as fictional personae, but as if they were actors capable of playing completely different roles in different series. Still, essential qualities of their personalities remained from role to role. The animator's crowning achievement, the series *Star Blazers*, used his ensemble of recurring animated "actors" in a blatant tribute to the rebirth of Japanese ethos and aesthetic, and the general notion of cultural rebirth. Again conquered by aliens, earth people have gone underground (into remission, so to speak). The earth has been ravaged by radiation, and the surface is uninhabitable. Earth gets a message from a princess on a faraway planet who says she can help the earthlings if they come to her planet for a new technology called, in fittingly techno-genetic lingo, Cosmo DNA.

Space Pirate Harlock—now he's called Wildstar—enlists the help of some kids, who refit the wreckage of the sunken WWII Japanese battleship *Yamato* (there was indeed such a ship) to the needs of space travel. At the end of the first episode, the resurrected steel hulk emerges from its grave and sets off for the princess and her new DNA formula. We learn that in one year, unless our heroes avert the disaster, radiation will seep through the crust of the earth, killing any remaining Earth people. Before each week's closing credits, a number appears on the screen indicating how many more days are left until the end of the world.

The show required a whole new concept of continuity, or discontinuity. On the one hand, the show is not episodic at all. Like a sci-fi soap opera, the plot can only be understood from episode to episode, and the story takes years to tell.

Characters and details set up in one broadcast may not "pay off" or even reappear until months later. In a nod to thematic continuity and recurrence, the word *Yamato* was the name for the Japanese empire back in A.D. 500. But even with all this attention to metacontinuous cultural and character-oriented details, the standard rules of continuity are broken as often as they are enforced. Several times the entire crew of the *Yamato* is killed in battle, only to reappear the next season as if nothing had happened. In the world of animé, characters, warships, and cultures need not stay dead. They simply recur, again and again, retaining their original data, or cultural DNA, with astonishing precision. And always, without exception, it is the children who recognize and implement the process.

The Japanese *Gundam* tradition, which eventually spawned our own Transformers and Power Rangers crazes, expressed this complicated relationship among technology, evolution, and children as well as it has yet been done. And, not surprisingly, only our children seem to get it.

Gundam look like robots, but they are actually "mobile suits"—giant robot/vehicles that contain their human operator. Most of the shows employing gundam explain that the robots were originally used in the construction of space colonies. When war broke out between the colonists and the earthbound imperialists, the gundam were refitted with weaponry and used instead as war machines. The gundam move slowly but with great force; their movements are choreographed much like the movements of traditional Shogun swordsmen—so much so that the first of the mobile suit shows was called *Shogun Warriors*. The most futuristic concepts of space colonization and technological evolution were, from the very beginning of the gundam tradition, married to the oldest and most glorious set of Japanese cultural references.

The true gundam explosion began in the early 1980s with a

series called *Mobile Suit Gundam* that very elaborately explored the themes of technological and human evolution. In the first extensive seventy-episode series, Earth has already built a vast number of space colonies on the Lagrange points of our planet (the distances from the earth's surface that allow a satellite to stay in a perfect orbit). A fascistic familial regime living on the colonies called the Jion (they salute one another with "Sieg Jion!") has concluded that earthlings are evolutionarily backward; the Jion want to take over the colonies and claim their independence.

Meanwhile, many children born on the space colonies have mutated. They exhibit psychic abilities as well as an innate understanding of how to operate the gundam suits. When one of these children, called "new types," wears a gundam suit, he can experience the device as if it were an extension of his own body. New types somehow sense the space around themselves in the suits, and can operate all the controls without any formal training.

The plot of the series deals with the relationship between this biological evolution of children and the human-directed evolution of technology. Each side in the war spends most of its energy coming up with new prototypes for mobile suits in the belief that the best gundam will win the war. But the new type children are not interested in the war or its noble causes. They simply want to test out new gundam prototypes. When these kids risk their lives, it's to sneak into the research and development plant of the enemy in order to play with their new toys. War only serves to create better technology with which the new types can test and develop their psychic powers. There are no more agendas.

Of course, the gundam series' producers do not salt their plots with all this technological innovation for purely philosophical reasons. The shows are sponsored by toy companies who hope to sell as many products as possible. Each new mobile suit prototype is also a new plastic model for fans to buy and build. Every week a program is on the air, a new gundam prototype appears on the shelves of toy stores. Further, by creating child characters who have an extrasensory experience inside the suits, the toymaker/TV producers are attempting to convince children that they can really feel what it's like to move, strike, and *be* like a powerful mobile suit gundam.

The success of the gundam toymakers in marketing their products, however, does not negate the willingness with which their audiences accepted the ideas and themes implicit in these marketing ploys. In fact, the speed with which these toys were developed itself constituted a sort of turbulence. There were so many toys, accessories, plot points, and characters being created, modified, and reintroduced individually and in relationship to one another, that it's hard to say just who, if anyone, was responsible for the overall cultural agenda of the programs and products. All we can say for sure is that kids delighted in programs and toys that married the concepts of technological evolution with human evolution, and that cast the children, themselves, as the "new types"— armed with psychic abilities and a unique, instinctual understanding of technology.

Ironically but not at all coincidentally, these programs themselves are driven as much by technology as they are by any personal visions of their adult scriptwriters. The shows and their themes are wrapped around technological innovation—toy robots with movable parts. The evolving features of these high-tech dolls advance the stories and concepts in this otherwise market-driven cosmology. Therefore, whether they realize it or not, the gundam marketers are merely reacting to the advancements in their own technology and the appetite of their young viewers. Because their viewers are children of chaos, these two forces turn out to be immensely compatible, and their marriage is depicted in the stories themselves.

Gundamnation

As the toymakers developed more and more versions of the gundam robots, they began to put hinges and exchangeable parts everywhere. An arm could turn from a laser gun into a spear and then into a searchlight. Legs could become wheels or wings. These robot parts looked like the wheels of trucks, the wings of airplanes, and the fins of rocket ships. The self-similarity of mechanical technology took gundam toys to their next evolutionary level. Why not make robots that could turn themselves into rocket ships or racing cars? And thus the transformers were born.

Robotech (1980) was the first of these "mecha" or "trans-former" shows to make it to the West. Hasbro, the G.I. Joe company, licensed the toys and sponsored a sequel program in the United States, the tremendously successful *Transformers* (1984). Since the Vietnam War, real-world soldier toys had been frowned upon for condoning institutional violence. The Transformer toys gave G.I. Joe a new market and a new, futur-istic image.

The children watching these shows are not reveling in the violence—most of the destruction is done to or by machines, anyway. The joy comes from a vicarious experience of power. In these programs, without fail, the children understand tech-nology better than the adults who designed it. According to the internal logic of the gundam/animé universe, kids have evolved so that they can exercise control over and friendship with technology.

This is why these programs are so threatening to adults and why *Power Rangers* (1993) has been banned in several countries and its products outlawed in a few U.S. counties. The fighting itself is highly stylized. Kids who today imitate the high-kicking choreography of *Power Rangers* would have been imitating the Green Hornet or Joe Louis in past decades. Kids, by nature, like to jump around and knock things over.

But whether or not *Power Rangers* or any other media can make kids violent, the show intimidates adults because it exalts the ability of children to exploit technology and recognizes their roles as evolutionary agents. The violence is a red herring.

The Power Rangers descend from two Japanese television traditions: the gundam prototypes and the "rubber suit" monster shows. The gundam tradition, as we've seen, was developed around particular products and toy innovations. The rubber suit tradition was equally dependent on real-world technology for its stories and structure, and gave rise to themes just as resonant. The limiting factor in the rubber suit shows was not the hinge-capacity of toys, but the budget for production.

The first of the major shows was called *Ultraman* (late 1960s) about a giant space humanoid who defended Earth against monsters (rubber-suited stunt men). Ultraman himself was clad in a simple red-and-silver space suit and helmet, while the enemies were almost satirically realized versions of Godzilla-type monsters. The climactic scenes of giant monsters destroying buildings were the most expensive to produce, so the *Ultraman* writers developed a few conventions to limit the need for the construction of models and the filming of pyrotechnics. They decided that because Ultraman was from outer space, he could only survive in the earth's atmosphere for periods of three minutes or less. The monsters who appeared on the earth always started out about the size of human beings, but grew suddenly into giants just before Ultraman's arrival (much like Barney, who grows from a stuffed animal into a man-sized dinosaur when summoned by his young friends). Each episode's spectacular climax featured a giant Ultraman fighting giant monsters and ripping apart huge sets, but was limited to three minutes' worth of footage.

Like Gigantor, Ultraman was controled from afar by children. The next stage of rubber suit shows was to include children in the action. Literally dozens of programs arose where a group of five children, in colored space suits, fought monsters much in the way that Ultraman did. By some convention or other, the five rainbow kids—always four boys in blue, yellow, black, and red, and one girl in pink—would merge into a giant and fight the rubber-suited monster, who had also turned into a giant.

Power Rangers combined the gundam and rubber suit traditions into a single series, and then added some production-motivated themes of its own, resulting in the TV show most fully, if accidentally, directed to the children of chaos. The Power Rangers are rainbow kids, who, when they are confronted by a giant rubber-suited monster, use transformer technology to combine into a giant gundam robot for the climax of each show. Although each of the kids has tremendous martial arts skills on his or her own, only when they combine forces are their magical powers fully realized.

During most of the action, the Rangers fight generic opponents called "putties." Basically human except for being faceless, the putties all look the same, and descend in great numbers. Like the progressively more dangerous opponents in a video game, the putties come in increasingly skilled varieties: putties, super putties, and Z-putties, named for their master, Lord Zed, who is the Rangers' mortal enemy. In each episode, Lord Zed eventually creates another rubber-suited monster for the Rangers to combat. Zed himself appears to be a casualty of technology—almost like Darth Vader, who, in spite of his terrific powers, is biologically dependent on a complicated breathing apparatus in his mask. Zed sits on a throne within a fortress on the moon, connected to a huge apparatus by clear tubes that process and then pump some sort of blood through his body. For Zed, who attempts to use technology as a means of control and domination, his fortress has become a prison, with life-support machinery as shackles.

Despite his apparent handicaps, Zed has the ability to create tailor-made monsters and then grow them into giants at the climax of the show. The monsters themselves are almost always a dinosaur of one kind or another. In this sense, the gifted "new types" who fight them battle against their own evolutionary past. When they come together at the end of each episode to assemble their communal gundam fighting machine, each of the Power Rangers calls upon a different dinosaur spirit. By shouting his own special slogan, each Power Ranger summons forth a giant mechanical dinosaur, which transforms itself into a component part of the huge, collective gundam robot. This transformation looks almost like the animated film footage depicting the evolution of monkeys, step by step, into human beings.

The Rangers then find themselves within the huge robot, and battle the giant rubber-suited dinosaur, samurai-style, from behind the robot's control console. The children thus coordinate their futuristic supernatural skills in order to combat reptilian enemies from prehistory. The gundam will only function if all the Power Rangers are participating together. The battle, then, is between our evolutionary future—the combined efforts of a rainbow of children—and our evolutionary past: the efforts of a single, techno-imprisoned dictator to maintain personal control over our planet.

The show succumbs to yet another level of technological determinism in its American incarnation, produced by Saban Entertainment. In a dubbed translation almost as removed from its original dialogue as Woody Allen's *What's Up Tiger Lily?* is from the movie it satirized, the American version of *Mighty Morphin' Power Rangers* exploits the fact that its child heroes wear lip-obscuring helmets and replaces their lines with entirely new ones. Scenes where the Japanese characters appear out of costume—the less expensive, noncombat scenes—have been completely rewritten and replaced with new ones using young American actors.

On a superficial level, this has given the producers an

opportunity to instill the program with some of their more baby-boomer era concerns. Although in Japan only the pink ranger is a girl, in the United States both the pink and yellow rangers are girls, leading children on the Internet to point out that in the combat scenes (from the Japanese footage) only the pink ranger has breasts protruding through her suit, while the yellow ranger is clearly a muscular boy. In an effort to appear racially unbiased, the U.S. producers cast three white Americans, one Asian, and one African-American—but a bit too literally assigned. In the first season the yellow ranger was Asian and the black ranger was African-American! Too many kids noticed the obvious coding so, in subsequent seasons, the colors were reversed.

However much the producers edit the show to conform with their ever-changing notions of political correctness, they still base their plots on what they see in the shows and toys as they come from Japan. This has led to some interesting and probably unintentionally resonant results. The plots conform to the particular style of monster that appears each week, which is, in turn, based on whatever toy monsters have been marketed in Japan the season before, which are, in turn, based on the almost random combination of component toy parts and molds. This tends to make the stories extremely self-reflexive. For instance, in one episode, the monster at the end looks something like a hyena. In a writing process something like dream interpretation, the American writers decided to make this a story about laughing and shame.

Terribly deep? Maybe not. But the plot elements of these stories don't just disappear at the end of each episode. The overall plot of the show lasts from season to season. Ask any eight-year-old, and he'll be able to tell you the two-year history of the white ranger, a reincarnation of the green ranger who, unlike the other rangers, originally got his powers from Lord Zed's minions and can have his power drained without notice. The metacontinuity of the story rivals that of an Asimov trilogy. Most important, these plot elements are all developed around existing footage that is itself developed around toy marketing.

This reverse-engineering of stories makes fully conscious control of the program's themes utterly impossible. Unintentionally, perhaps, but undeniably, the overriding theme of the show turns out to be co-evolution with technology. In addition to the way the Power Rangers use technology to fight monsters of the tyrannous past, they depend on technology for moral and spiritual guidance. When the Power Rangers are in trouble, they turn to Zordon, a disembodied ageless sage who acts as a techno-oracle from within a computer. Pure consciousness available only through a communications device, Zordon demonstrates that the wisdom of the ages has the ability to speak to us through technology, just as the Power Rangers show how children may be able to bring us into our evolutionary future via the same means.

The themes embodied by the Power Rangers have not gone unnoticed by other, more traditional cultural institutions. An outspoken proponent of many of the series' agendas, House Speaker Newt Gingrich invited the Power Rangers to be his highly visible guests at the first session of his new Congress in 1994. This was not a random grab at popular culture. More than almost anyone in politics, Gingrich is aware of the power of media imagery to influence the public's perception of policy-making. His belief that mediating technologies can evolve democracy in America to the next level is embodied by the sci-fi rainbow heroes, and is much more accessible than Al Gore's "information superhighway" language and aesthetic. Claiming the futurist turf of Alvin Toffler as his home territory, Gingrich proposes we embrace the possibilities for personal empowerment offered by the computer, even going so far as to offer a tax credit to low-income families for buying personal computers. Like his mentor Toffler, Gingrich believes that technology can serve to restore a natural balance of power between people and their government instead of perpetuating a system where citizens feel disconnected from their representatives' decision-making.

It is difficult, however, for most Americans to accept the notion that technology can express rather than stifle the course of nature. Perhaps the most presumptuous of the many polarizations spawned by the dualistic worldview, the idea that technology and nature are separate and at odds, presupposes that humankind is capable of creating something beyond the scope of nature. After all, what isn't natural? Our buildings, chemicals, and machines? What about beaver dams, bee's honey, and lobster claws? If humans are understood as part of nature, then the things we create are extensions of ourselves, and extensions of it. Amazingly, it was the Japanese, more victimized by technology than any other culture in history, who in their sci-fi media came up with a coevolutionary model of human beings and technology.

Still, Japanese storytellers understand that adult consensus culture will not accept this co-evolutionary model without a fight. In the animated film and comic book series *Akira*, both cult hits, children of postapocalyptic Tokyo are slowly mutating into new types with psychic and telekinetic abilities. Although the army and government struggle to keep these children imprisoned, the adults are fighting a losing battle. The more they attempt to repress these emerging talents the more violently they finally erupt.

The organic abilities of these special children only anticipate abilities we are all gaining in real life through technology, anyway. The adults in *Akira* fear children who can read their thoughts. The loss of privacy leads to tremendous paranoia and extreme measures are taken to maintain secrecy. Isn't this same phenomenon occurring now in electronic mail and online cash transactions? One of the biggest issues in communications today is whether we can keep information private and finances secure. And from whom are we keeping it secure? Hacker kids, who seem to have an uncanny ability to slip through the cracks of our security programs.

The freedom of expression offered by the safety of the animated media has also yielded some other disturbing if predictable results. Animé kids, particularly the girls, are getting increasingly sexy. Ranging in age from about nine to nineteen, the animé girls are often only part human. As designer beings, they are physically "idealized," commonly depicted in skintight plastic jumpsuits. College boys in the United States and Japan collect posters, videos, and models of the perfectly proportioned teenage heroines and exchange graphics files of their favorite images over the Internet. While the most avid fans at Japanese university animé clubs assure me that the youngest characters (under ten years of age, to be exact) are rarely the objects of their sexual fantasies, the resemblance of animé girls to kiddie porn is somewhat unsettling, especially to Western audiences. But the Japanese, more desperate for a vision of the future after nuclear destruction, embraced the idea of evolutionarily advanced child cyborgs who could defeat monstrous abusers of technology, which in turn may have helped them overlook some of the sexual overtones of their programs.

Whether or not its aesthetic and tools are used to create idealized young female forms, cyber is still pretty sexy all by itself. Mediating technologies, be they telephones or "teledildonically" equipped cybersex virtual reality suits, forge intimacy between people. Though masked by the anonymity and distance of electronics, many social interactions occur between people who would not have interacted otherwise, thanks to technology. Still, it's pretty easy to judge those who get intimate through or with technology as socially retarded. It could certainly be argued that the fascination of college kids with cartoon babes is based in a fear of intimacy. Cartoons don't demand anything from you. They are even less reality-based than an airbrushed *Playboy* centerfold and present a picture of objectified, streamlined, and designer femininity. They are not real girls, so they provide a more guilt-free experience of kiddie porn than the exploitational photos taken by child molesters. The kinds of young men lusting after the animé girls, however, are not generally the same sort who frequent porn theaters or buy obscure and deviant sex magazines. These are *Star Trek* fans and engineering students who, while stereotypically shy with real-life women, also muse on the possibilities depicted in the science fiction of Philip K. Dick (in the movie *Blade Runner*) or William Gibson (*Neuromancer*). Although the fans may fantasize about programming the ultimate sexual experience in the virtual reality "holosuites" on *Deep Space Nine*, their sexual attraction to animé girls is more than neoprene-deep.

When a group of children or adolescents are linked to one another through bizarre genetic mutation, psychic fields, and electronic transceivers, their kinship as new types can't help but take on a sexual tone. Sex, after all, is the most natural method of genetic engineering and evolutionary foresight. The children of animé films are forced to take on adult roles as they work to save their city or world from apocalypse. Our attraction to them is more than mere pedophilia—it is a yearning for a continuation of our species. It is a genetic attraction as much as a physical one.

Maybe because of our Puritan heritage, it has traditionally been more difficult for Westerners to appreciate children directly. Just as we are afraid to expose our children to violence, we are afraid to expose ourselves to their emerging sexuality. We like to think of children as ignorant and innocent. When they reveal to us that they are hardly ignorant, we must accept that they are hardly innocent, either. Accordingly, America handles the potentially precocious children of its futuristic media with kid gloves. We also tend to hide with equal determination from the themes they and their media embody. Still, however well we gloss over what we fear are the more culturally dangerous of our sci-fi media's implications, even more threatening concepts manage to shine through. For sex, like violence, is pretty easy to recognize in media, and similarly easy to combat. Concepts like chaos are a bit more slippery.

Generation Treks

Having emerged from World War II victorious and relatively unscathed, we were less compelled, at first, than the Japanese to explain the devastating effects of technology. The imperialist paradigm still worked for us pretty well, and we were not hard-pressed to rationalize the potentially negative effects of our inventions with a coevolutionary model of the future. Instead, we were left with the lasting impression that our tech was better than everyone else's tech, and understood that as long as we kept things that way, we could run the show.

Our science fiction media, as a result, didn't assume the task of envisioning a postapocalyptic world of the future, where the human race somehow evolves toward a sustainable technobiological ecology. Instead, sci-fi used the alien aesthetic of an imagined future as a backdrop against which contemporary dilemmas could be observed in great relief. The ongoing development of science fiction media became a chronicle of cultural history in the present.

Take *Lost in Space*, the mid-1960s sci-fi *Swiss Family Robinson* adventure brought to us by disaster-meister Irwin Allen (*Poseidon Adventure, Towering Inferno*). The show did not work on the level of science fiction or even adventure. Instead, the experience of the late-twentieth-century Swiss Family Robinsons, alone on an uncharted planet, recapitulated the isolation of American life on the newly developed suburban frontier.

The experience of growing up in the 1960s and 1970s

divorcescape was different from what the community planners had hoped for. Tract housing and competitive shrubbery did not lend themselves to strong bonding between families. Instead, each nuclear family holed up in its own aluminum-clad, fiberglass-insulated, and soundproofed split-level colonial and conducted its business in private. Sure, we each had our own backyard and unfinished basement, but we were lost in all that space.

Will and Penny Robinson were similarly removed from the company of other boys and girls, and sentenced to a decade or more in the *Jupiter 2* with their mom and dad. Like the rest of us, the Robinsons went off course and wound up lost in space. Shipwrecked on a lonely planet at the end of their first season, they spent the next two years hosting any aliens who happened to arrive on their grounded saucer's doorstep. Like those of us back on Earth for whom neighbors had become competitors or even threats, the Robinsons needed to regard each new visitor with a wary eye and a laser gun. Even in this show, the children had a better sense than the adults of the aliens' true inner natures. Young Will was always the one who recognized, before everyone else, when an alien was dangerous, while pretty Penny (herself played by a *Sound of Music* veteran) could see through to the warm heart beneath the reptilian shell of the ugly aliens whom everyone else feared unnecessarily.

If it weren't for the strength of supermom Maureen and her ability to maintain the cohesiveness of the nuclear family, the Robinsons would have had no social institutions to guide them whatsoever. When June Lockhart scanned the alien horizon (with the same moist-eyed concern she employed in the *Lassie* series back on 1950s' Earth) for whoever hadn't made it back to the ship by dusk, latchkey kids across the nation got a vicarious taste of the enduring quality of a loving, stay-at-home mother. Will Robinson's only consistent comfort, other than the bosom of mom, was technology: the Robot. Like the

85

flickering TV tube through which we collectively celebrated our suburban angst, the Robot's flashing-red voice-screen provided the boy with the closest thing to a best friend this side of Alpha Centauri. When the world no longer provides a tenable social scheme, technology rises to the occasion. That is, technology and Dr. Smith, whose antics with the pubescent boy rivaled Batman's with Robin for its homoerotic overtones. Moreover, their relationship, mediated by the robot, intimated that technology may be the only link left between science-bound youth embodied by Will, and the history of culture, in the portrayal of the elderly Dr. Smith by Shakespearean-tongued actor Jonathan Harris.

Lost in Space typified, more cartoonishly but also more obviously than most other sci-fi series, the way futuristic media explores and exploits the social tensions of the present moment. Similarly, the near thirty-year history of *Star Trek* media has never had anything to do with our futures—except for those of us lucky enough to have invested in Paramount or Viacom at the right time. *Star Trek*, the original series, was one of the most topical programs of all time. The issues of the day, like reckoning with hippies or the Civil Rights movement, found their expression in simple, episode-length metaphors. "But can't you see, he's black on the left side of his face, and I'm black on the right," explained a two-colored humanoid attempting to rationalize a centuries-old race war in a distant galaxy. Although naïvely direct and pathetically moralistic in hindsight, this series made waves (and a certain sort of progress) in its day by letting white Captain Kirk kiss his black communications officer (starship-speak for receptionist) Lieutenant Uhura, even though they were forced into the embrace against their will by telekinetic aliens.

Captain Kirk's *Star Trek* was produced by and for pre-boomers—those World War II–era folks who were confused by our nation's waning enthusiasm for Vietnam, festering guilt over racial inequities, and increasing paranoia about a Cold

War nuclear threat. The *Enterprise* looked and sounded like a battleship, and issues were laid to rest through physical confrontation, whether Kirk fought hand-to-hand against a Romulan or shot photon torpedos at a Klingon Bird of Prey. The future was used simply to reassure audiences that our dualistic, balance-of-power paradigm would work for centuries to come. The only time children were permitted into the plots was to demonstrate just how helpless they were without adult supervision. Even though the kids on one episode, called "And the Children Shall Lead," had psychic, new-type abilities, the powers were useless, even dangerous, because the kids had no moral template with which to direct their evolutionary skills.

The baby-boomers had their turn with the aptly titled *The Next Generation*. A bald-is-beautiful diplomatic captain (played by an RSC veteran) at the helm of a kinder, gentler *Enterprise* negotiated his way through crises with the help of his "empath" counselor/psychologist. Science fiction's *Big Chill*, *TNG*'s sensitive crew utilized New Age philosophy and the tao of physics to navigate through the relativistic haze of a politically correct galaxy. The first officer, Commander Riker, occasionally exhibited enough testosterone to beat someone up or bed an attractive alien, but he made up for it by consistently refusing offers to command his own ship. It's as if he knew he'd have to change his ways in order to grow up in the touchy-feely do-gooder Federation.

The Next Generation's treatment of youth, however, demonstrated a markedly increased awareness of the role children would need to play in our cyberspiritual future. Wesley Crusher, son of the ship's doctor, was a stereotypical nerd-genius in the first several seasons of the program. In one of the final episodes, though, Wesley quits the prestigious Starfleet Academy to become a shaman. He has realized that he has the ability to travel interdimensionally without the help of a warp drive engine or tachyon pulse, and vanishes into the ether to explore alternate realities and assist in the unfolding of time itself.

The next offering in the *Star Trek* saga was *Deep Space Nine*, which served as *Bonanza* to *TNG's Gunsmoke*. An intergalactic hitching post, *Deep Space Nine* is a space station positioned just at the entrance to a "wormhole"—an intergalactic portal to the other end of the universe: the final-est frontier. On one level, the wormhole convention reeks of writers' guidelines and pilot proposals, transparent as an all-too-convenient method of bringing new kinds of aliens to the program. On the level of cultural allegory, though, the discovery of the wormhole and the issues it raises mirror the dismantling of the Iron Curtain and felling of the Berlin Wall, which opened up the West to formerly isolated, alien societies of our world. The artificial duality between Eastern and Western Europe dissolved (not without creating other problems, of course) and Soviet bloc nations that did not even have phone lines to the West became our trading partners and media consumers. Anything could pass between us. Decades of work casting ideological distinctions between "us" and "them" were neutralized. Now that the floodgates were open, how were we to respond?

Thematically, then, *Deep Space Nine* appears designed to tackle these problems of our post–Cold War "new world order." Unlike previous *Star Trek* incarnations, we no longer journey through space in search of new planets and civilizations. The space station remains still. There is no new territory left to conquer: just wait for attacks and hold down the fort. Like a week on CNN, delicate alliances and negotiated neutral zones shape the world of *DS9*, while terrorists provide the main events.

DS9's stasis, however, is what provided the anchor for the first *Star Trek* series aimed directly at the concerns of the so-called Generation X. A billboard in San Francisco advertising the premiere of the program displayed an astronaut putting a flag on the moon—a photo evoking both MTV's logo and the first real-world effort of human beings to colonize space—but the caption read, "Been there, done that. What's next? *Voyager*." On one level, it's just a cute joke: We've been to the moon, now let's see what's farther out. It's also a dig at MTV: you've surfed channels and lived the MTV revolution, now watch some real *television*. Generationally, the ad makes an even more significant statement. GenXers have been told time and time again by their boomer-age parents that each new idea they have is just a rehash of a 1960s idea. A rave? Psychedelics? Coming together into a colonial organism? Planet consciousness? Been there, done that. *Star Trek Voyager*, at least according to the ad, will give this generation a chance to experience a little "been there, done that" themselves. The moon and American flag of their parents' day? Kids' stuff. *Voyager* will take you to uncharted turf.

The crew of the *Voyager* starship is the first to venture beyond the reach of Federation command. The mean, lean *Voyager* starship was on a mission to capture a renegade band of violent but essentially virtuous young freedom fighters known as the Maquis. During the chase, a creature called "The Caretaker"—an uninvolved alien from across the universe—kidnapped both crews in an attempt to save a planet for whose inhabitants he had become an overprotective parent. The Caretaker dies, leaving his surrogate children to perish and, worse, leaving both the *Voyager* and Maquis crews stranded seventy years warp-speed travel away from home—out of communications range with their own guardian institution, the United Federation of Planets. The formerly combative crews team up for their odyssey back home.

Despite its painfully politically correct casting of fuzzy and ethnically diverse crew members (a female captain, a Native American first officer, a black Vulcan) the show still captures much of the renegade spirit of its current baby-buster audience. Out of the range of Federation transmissions, the young crew must make its own decisions on a case-by-case, moment-to-moment basis. Cooperation is the key to succeeding in a world where the parents have all disappeared, leaving only a few sets of instructions behind them. The moral templates and social contracts of the past are dispensed with, as a crew of Starfleet dropouts and political defectors cast off their superegos and get down to business. Their only link to the past—like Will Robinson's in *Lost in Space*—is in the form of computer files and interactive programs. Even the ship's doctor—a character traditionally used on *Star Trek* shows as the voice of human morality and compassion—has been replaced by a computer-simulated holographic physician who, though programmed with all known medical knowledge, is utterly detached from A.M.A. patriarchy. The *Voyager* series serves its mostly twentysomething audience by recapitulating the GenX conundrum against the stark but crystal clear backdrop of outer space.

Voyager, not coincidentally, is the first of the *Star Trek* series to be conceived completely after creator Gene Roddenberry's death. Left on their own, the producers wrote a show mirroring their own experience of independent choice. Their decision to free the characters from their boomer quagmire did not constitute a rejection of the past any more than the GenX choice to reevaluate some of the more longstanding cultural assumptions of its own world constitute a blanket rejection of human history. It's just evolution.

But on *Star Trek* evolution has its limits. Reasonable distrust prevails. Since the crew is stranded in space, many episodes take on the quality of *Gilligan's Island*: A strange person or group appears with the ability to get everyone back home, but something happens to screw it all up, and the *Minnow*'s five passengers remain shipwrecked. In the *Star Trek* universe, the obstacle to negotiating passage home is not Gilligan's wacky mishaps, but old-paradigm Federation dualism. In one episode of *Voyager*, a renegade member of the crew has begun negotiating with an alien race in the hope of trading for technology to get home. The captain of the *Voyager* discovers these treasonous activities and immediately puts a stop to them.

The argument that ensued between captain and crew member sounded as if it were lifted from Jane Jacobs's highbrow treatise on commerce and politics, *Systems of Survival*. According to Jacobs, there are two main syndromes guiding international relationships: the commercial moral syndrome and the guardian moral syndrome. Commercial morality is employed by businesses and depends on continuing innovation and development of relationships: shun force, come to voluntary agreements, respect contracts, be optimistic and competitive. Anyone who hopes to do business with strangers must proceed with the assumption of competitive good faith. Guardian morality, on the other hand, employed by governments and not businesses, depends on secrecy and tradition: be obedient, respect hierarchy, take vengeance, be fatalistic and ostentatious. More than anything, governments strive to preserve the sanctity of their nation-states and the loyalty of their constituents. They need to maintain the status quo and, if possible, expand by conquering their neighbors. They do not offer profit and dividends to their shareholders; they offer security to their subjects.

When the captain of the *Voyager* argued against sharing technology with aliens, she was employing the guardian morality. She claimed that helping strangers could "upset the balance of power." Working against expediency and the laws of international commerce, she opted instead for the maintenance of secrecy and discrete societal boundaries. Her crew remained stranded in space, and the attempt at commerce was considered treason. No matter how GenX its pretensions, *Voyager* is not ready for libertarian-style chaos.

The entire *Star Trek* saga deals with technology and culture from a dualistic perspective. The "prime directive" that all

Starfleet officers must swear to uphold is one of "noninterference." No alien culture is allowed to be altered by their observa-
tion. But how is this even possible? Quantum physics shows us
that nothing can be observed without being changed. Moreover,
the noninterference oath presupposes that there is some difference between human activity and the rest of nature. In an anti-
imperialist frenzy, Gene Roddenberry and the creators of *Star
Trek* could imagine nothing between total domination of a culture and complete separation. They apply the same rules of
noninterference to time travel: you can go back in time, but just
don't change anything important or the future will be different.
Again, by the laws of chaos mathematics (feedback and itera-
tion), a tiny change—even moving a matchbook three inches—
can lead to huge, systemwide repercussions once enough time
has passed. The effort to maintain the established order of
things is, at best, futile. At its worst, it actively discourages the
process of nature.

Other, more recently conceived, American science fiction
shows like *Babylon 5* and *Earth 2* (notice how many programs use numbers in their titles—*DS9*, *Earth 2*, *Babylon 5*,
the virtual reality show *VR5*—accepting themselves as prototypes, or evolutionary levels of their own genres) suffer a bit
less from the guardian paradigm, mostly out of necessity.
Babylon 5 is the last of a series of space stations developed to
promote peace between Earth and alien cultures, like an
interplanetary United Nations. While mistrust and careful
diplomacy are the operating principles of the show, the plot
and world of interactions is so large that it creates a very different style of experience for the viewer. Like the classic science fiction trilogies of Asimov or Bradbury, *Babylon 5* is
epic in scope. Plot points that arise and characters who
appear in one episode may not be resolved until seasons
later. Characters are often played by heroes from other science fiction programs—like Bill Mumy, who played Will
Robinson on *Lost in Space*. When a show like *Babylon 5* can

92

create an entire Marvel Universe of its own it not only guarantees a devoted following, but also frees its stories from achieving premature closure, or spelling out conflicts too succinctly. Unlike *Star Trek*, *Babylon 5* is not a black-and-white universe of prohuman and antihuman aliens, but a myriad of gray alliances, double crosses, and surprising twists. A villain one day can be revealed as a hero months later. Because it is not a strictly episodic show, the situation does not need to return to a state of order every week. The show is so complex that fans distribute lengthy glossaries, plot summaries, and character lists over the Internet.

But however intricate and chaotic a universe it intimates, *Babylon* 5, like most American science fiction, makes the unstated assumption that no matter how far technology evolves, human nature will remain basically the same.

Technology and humanity are understood as two distinctly different lines of development, in strict conformity with the guardian moral syndrome. The roles of human beings and the structure and function of society are constants. Although technology may be evolving, as long as it remains controlled by the same forces nothing fundamental needs to change.

Even the visionaries of the American cyberpunk genre, however forward-thinking, refuse to consider the possibility that people may develop along with their technology. The worlds of cyberpunk authors like William Gibson and Bruce Sterling, though stunningly dense with inventive technology, are largely devoid of love and spirit. The children have no psychic abilities, unless they attach "dermatrodes" to their skulls or are participants in experiments funded by evil governments or terrorists. The stories revolve around black markets in human organs, drug-addicted computer hackers, and supremely powerful corporate executives. The technologies of the future simply serve to promote further the innate, destructive expression of the human will. The message of this genre is that the more things change, the more they stay the same. Human nature is awful, and no matter how fancy our technology gets, our basic nature will remain as it is.

In both cases—the optimistic, pro-American vision of *Star Trek* and the pessimistic anticorporate vision of the cyberpunk authors—human evolution is conspicuously absent. Occasionally there is an accidental or experimental mutation, but it always yields an ultimately inferior being. (The superman genius played by Ricardo Montalban in *Star Trek 2: The Wrath of Khan* was the result of genetic experimentation, but his superior strength and intelligence turned him into an evil madman.) No, biological and cultural evolution are either things to be avoided or impossible to imagine.

The eagerness of the Japanese to embrace concepts that have been eschewed by American science fiction is not a function of aesthetics but necessity, still the truest mother of invention. The Japanese learned through experience that the evolution of technology without an accompanying development in human interaction can only end in disaster. This is not to say that human beings are intrinsically evil—simply that technology can, like a feedback loop, amplify and dangerously exaggerate some of our social shortcomings. The Japanese animé notion that we can biologically evolve our social and communicative abilities is replaced in the patriotic *Star Trek* universe with the assumption that human beings must strive toward behavioral perfection by developing complex sets of rules and following them exactly. The underlying logic of the Western vision is Christian fundamentalism: Created in God's image, the human being is itself a perfect form and can only be sullied by sex or corrupted by technology. Therefore we maintain a strict division between the ideas of technological development and the sanctity of humanity. Whether it's the traditional science fiction writers who pit the ideal human being against the debilitating side effects of his technologies, or the "new edge" cyberpunks who arm the chronically sinful human with morally neutral technologies, people and machines just don't mix. You can't improve on nature. People will be people.

Only in the Branch-Davidian-Sarin-gas-in-the-subway 1990s do Westerners begin to feel the crush of apocalyptic thinking and rush to develop new visions of the future that include a little evolution for ourselves. To hell with perfection; let's just soldier on past the turn of the century. Amblin Entertainment's *Earth 2* takes place after our own apocalypse, on a planet that a few dozen settlers hope will serve as the next Earth. Perhaps out of necessity, the characters on this show seem to accept that their tools are not the only things that change over time. Their only goal is to survive, and like the original American pioneers on the Oregon Trail, the *Earth 2* settlers travel across the unfamiliar terrain in covered wagons, depending on the knowledge of the new territory's own natives, the mysterious, tribal Terrans, for wisdom of the planet's nature. The Terrans are a deeply spiritual race who communicate through telepathy. They have the ability to live within the skin of the planet's crust, and offer guidance to the humans who have not yet learned how to live in harmony with their environment.

Strikingly unique to this program, whose 7:00 P.M. time slot marks it as family entertainment, is that the humans themselves are evolving new abilities. One young man can communicate with the Terrans through his dreams. The Terrans also saved the life of one of the settlers' children by genetically altering him. Displaying some remarkable new-type abilities of his own, the ten-year-old child was able to negotiate the amnesty for a condemned segment of the Terran race. The Terran abilities—psychic awareness, group mind, aboriginal technology—are the same survival skills that the settlers and, it is hoped, their audiences realize are essential to develop in themselves if they are going to stand a chance of surviving in the age of chaos.

Gump Fiction

While sci-fi media intimates the possibility of coevolution with technology as an alternative to apocalypse, the broader implications of such development can only be fought out in mainstream popular culture. The most potent and unsettlingly convincing ideas of the screenagers have naturally trickled up from kids and fringe media to film and television at large. This process, in itself, poses a threat to the seeming sanctity of status quo ethics and values.

The way media tends to promote the coevolutionary agenda exposes our two main psychological obstacles to its almost inevitable rise: fear of fragmentation and fear of monoculture. Every conversation about a global technologically mediated society eventually falls into one of these two nightmare scenarios. The fragmentophobes worry that with hundreds of cable channels and thousands of computer conferences, people will be able to isolate themselves within their own extremely opinionated factions, never being exposed to new or opposing viewpoints. The fragmented culture will be disjointed and incomprehensibly chaotic. In worst-case scenarios it would lead to heavily armed survivalist cults throughout the world. Tolerance would give way to fundamentalism, and our cooperative society of competing special interests would polarize into a war among irreconcilable antagonists.

The monocultural vision prophesies that powerful corporate and governmental forces will hijack the information superhighway and use it to program our voting and purchasing decisions. They fear that a Rupert Murdoch or Viacom will gain a controlling interest in the global information networks and turn them to their own ends. When we're all plugged in, we will all be susceptible to mind-numbing brainwashing efforts. Intellectual elitists without Big Brother paranoia opine that the same monocultural whitewash will occur even without any intervention from the world's power centers. Give the people what they want, argue the overeducated, and they will choose what's worst for them. Using ugly words like "masses," the intellectuals complain that without proper guidance, the technologically liberated population will attempt to construct a culture out of Twinkies and *America's Most Wanted*. Again, intolerance will rear its ugly head, and intellectuals who disagree with the masses will be hanged, or at the very least have their university tenures revoked. The horror.

Anyone who understands or, better, has experienced being a member of a networked community of individuals knows that neither of these nightmare scenarios will come to pass. The fear of fragmentation is a fear that unfettered self-determination would act as a centrifuge on society, leading each of us in turn to his own cable channel or Internet chat. But a tour of the remote regions of the cable box or Usenet group list reveals that the same images, people, and ideas seep from one channel or group to another. A networked mediaspace keeps each of us intimately involved in one another's content. The footage shown on MTV might contain images from CNN that themselves are related to those on *Roseanne* and Coors beer commercials. When Viacom owns MTV and Paramount and Simon and Schuster, a franchise like Beavis and Butthead will show up on the music channel, *Hard Copy*, and in trade paperback books. There's no escape.

Precisely the concern of the monoculturephobes. What they don't realize, however, is that turbulence makes central control of a culture impossible. In its acceleration down the information superhighway, the cart has overtaken the horse, making the modern mediascape utterly undirectable. There is simply too much media, changing too rapidly for anyone to control it. The media conglomerates are at the mercy of the market forces they used to stoke. Rupert Murdoch's business (which paid for this very book) has just one bottom-line motivation: to make money. In order to do so, his only choice is to feed human hunger. If our appetite is for *Friday the 13th: Part 26*, then that's what we'll get. If our appetite is for a mediated cultural organism, then that's where we'll direct our dollars and, accordingly, Murdoch's efforts.

This is why the biggest deals on the front pages of the *Wall Street Journal* and *New York Times* business section are no longer about advertising contracts or the purchase of film libraries, but the construction of conduit. Last century, people wanted ways to get their bodies and merchandise to and from the West. The men who got rich off this desire were the railroad barons who contructed the conduit. Today our culture is asking for lines of interconnectivity and ways to exchange their ideas and information with others from around the world. Accordingly, the people who will get rich in the twenty-first century will be the ones who build the digital infrastructure. However much they get paid, they are still in our service.

If the quality of our collective appetite leaves something to be desired, so be it. We will get what we ask for. Luckily, the truly unfettered appetite generally leads a hungry person to food that is right for him. Experiments conducted on babies demonstrate that if given the opportunity to eat any food at all, a baby will, over a two-week period, consume a fairly balanced diet. An entire day might be spent eating candy, but the next day the baby may consume only lettuce, and so on. Babies naturally, eventually, eat what's good for them.

According to the paranoid elitists, masses are to be feared also for their intolerance of diversity. If market forces rule, then classic forms like ballet and opera will disappear, not to mention tribal cultures or nonelectronic media like books. But at what moment are we to freeze-dry and preserve such cultural strains? Aren't they also the result of centuries of evolution and hybridization? Should we have outlawed the Spenserian sonnet for replacing the Italian ottava rima, the shoe for replacing the sandal, or Windows for replacing DOS?

The best elitist arguments emphasize the domination and

absorption of one culture by another. The Conquistadors vanquished the Aztecs, the Dutch herded the Zulu, and the Americans decimated the indigenous populations and enslaved Africans. But a genetic model of civilization would force us to accept that conquering a people is not the same as permanently wiping out their culture. Genetics shows us that genes don't just disappear, but are retained from generation to generation; the oldest recessive genes can reappear as biological traits generations later. Unless they affect survivability, the relative proportions of genes in a gene pool remain constant. They will recur. It follows that the only way to destroy a society's cultural genes, or what have become known as "memes," would be to kill all the people who carry them.

For absorption is a two-way street. When African slaves were brought to America and divorced from their homeland, their own cultural identity was not completely neutralized, however inhumane the holocaust that was perpetrated against them. The African Americans were no more "Americanized" by their captors than was American culture Africanized by the slaves they took. Gospel, rock and roll, the Broadway chorus, not to mention rap music and "phat" clothes, are all strains from African culture. Likewise, Christianity no more overwhelmed the pagan traditions of Europe than the pagan traditions infiltrated Christianity. The Christmas tree is itself derived from an old German solstice ritual, and many saints are based on pagan gods. Thanks to the media infrastructure and its ability to accelerate the interchange of culture and percolation of buried ideas, we'll be seeing a lot of formerly "recessive" memes emerge in the coming years. The particular cultures responsible for their original inception may lose credit, but that's not the point. The cultural memes survive.

If anything, the maintenance of heightened vigilance is the enemy here. If we are too afraid to participate in the burgeoning global-tech culture, we will prevent it from reaching necessary turbulence. It's like riding a bicycle—go too slow and the wheels won't spin fast enough to generate angular momentum. You'll lose your balance and fall off. Worrying about falling to the left is no different from worrying about falling to the right.

The choice is whether to stick to oversimplified dualistic morality and stunt the acceleration of discontinuity whenever possible or to celebrate turbulence and, in doing so, learn to navigate chaos. These two opposite reactions to the stress of living in a technologically enhanced culture are at the heart of the much-publicized but little understood generational rift between the baby-boomers and baby-busters. The famously "X" Generation is really just the first of the screenagers to come of age and express, to the culture at large, their acceptance of discontinuity and turbulence. The disjointedly random style of the cult film *Slackers* and the hybridized phraseology ("McJobs," "Bradyism") in the margins of the novel *Generation X* anticipate the chaotic worldview of the screenagers to follow.

The disparity between the boomer and buster reactions to chaos was beautifully embodied by two of the most successful films of the mid-1990s, *Forrest Gump* and *Pulp Fiction*. Both movies used technology, sampling, and discontinuity to present alternative pictures of our increasingly turbulent times. While screenage progenitor Quentin Tarantino's *Pulp Fiction* celebrated the potential of nonlinear storytelling to express the joy and confusion of postmodernity, boomer Robert Zemeckis's *Forrest Gump* celebrated the ability of media technology to rewrite and simplify history.

Zemeckis has a history of revisionist filmmaking. He uses special effects, as in movies like *Back to the Future*, to give characters a chance to go back in time and change the way things happened. The magic of technology—whether manifesting as the time-traveling DeLorean sportscar in the movie or the special effects used by the filmmakers to show us multiple Michael J. Foxes from different eras in the same frame—is used to resolve the ambiguities of the past before they can iterate into big problems for the future.

Forrest Gump, too, takes the form of a flashback—this time into the memories of a half-wit. We relive the most irreconcilably disjointed moments in baby-boomer history, using special effects to smooth over their ambiguity. The joy of seeing Gump pasted into famous footage and scenes of recent American history is that we are permitted to reevaluate the troublesome aspects of the last few decades from a simpleton's point of view. Gump's lack of awareness allows him to fall, by sheer luck, into good fortune at every turn. He becomes a war hero and multimillionaire by blindly stumbling through life with nothing more than the good morals his mom taught him, while the people around him who seem more aware of their circumstances drop like flies from war wounds, AIDS, and other disasters we're to believe probably wouldn't have happened to them had they only been as steadfastly faithful to basic morality as Gump. In this story's hopelessly dualistic schematic, Gump is saved and most everyone else is damned.

Forrest Gump uses tricks of film technology to smooth out the inconsistencies and unpredictability of modern life, allowing its audiences to maintain their static, postwar worldview in the face of the approaching chaos. The unpredictability of life is reinterpreted as a box of chocolates—"You never know what you're gonna get." But it's a box of chocolates, for heaven's sake! You can pretty well count on getting a chocolate as long as you don't reach outside of the box into the real world of sharp rocks and biting bugs. The movie urges us to stay inside the box. Just stay in the box, and you will be guaranteed a sweet, continuous, and linear experience. No big surprises. Just train yourself to be as unconscious as Gump. Don't think, reconsider, or question anything. Awareness will disconnect you from the random benefits of coincidental grace. The opening sequence of the movie tells it all: in one continuous shot a feather floats on the wind, effortlessly wandering over the rooftops of a small, perfect town, and lands at Gump's feet, either coincidentally or by divine will. Of course it was neither luck nor God guiding the feather's path, but the will of the movie's director, who used cinematic trickery to create the continuous sequence. Just like Gump, we, the audience, are kept ignorant of the special effects, edits, and superimpositions, as technology is exploited to make the facade look seamless and real. And what does Gump do with the feather? He puts it in a box with his other collected trinkets.

The contrast between *Forrest Gump* and *Pulp Fiction* reflects the disparity between the pre- and postchaotic worldviews. Where *Gump* offers us a linear, if rewritten, historical journey through the decades since World War II, *Pulp Fiction* compresses imagery from those same years into a stylistic pastiche. Every scene has elements from almost every decade—a 1950s car, a 1970s telephone, a 1940s-style suit, a 1990s retro nightclub—forcing the audience to give up its attachment to linear history and accept instead a vision of American culture as a compression of a multitude of eras, and those eras themselves being reducible to iconography as simple as a leather jacket or dance step. The narrative technique of the film also demands that its audience abandon the easy plot tracking offered by sequential storytelling. Scenes occur out of order and dead characters reappear. On one level we are confused; on another, we are made privy to new kinds of information and meaning. The reordering of sequential events allows us to relate formerly nonadjacent moments of the story to one another in ways we couldn't if they had been ordered in linear fashion. If we watch someone commit a murder in one scene, our confusion about his motivations can be answered by going backward in time in the very next scene. *Pulp Fiction* delights in its ability to play with time, and in doing so shows us the benefits of succumbing to chaos.

Quentin Tarantino's background as a video store clerk and

avid pop culture fan are as evident in his movie as the philosophy of *Back to the Future* is in Zemeckis's. As a screenage kid who could pause, rewind, and change tapes any time he liked to, Tarantino got used to appreciating footage as fodder. Pulp. Take any piece of it from anywhere in history and paste it in, wherever you want. The library of videos at Blockbuster is Tarantino's equivalent to the Marvel Universe. Characters can move from one film or genre into another. The object of the game is to avoid getting freaked out by the resulting gaps, juxtapositions, and discontinuity.

The plot, so to speak, of *Pulp Fiction* recapitulates this experience of chaos. In the movie's world, the overall goal is stay cool in desperate situations. People who panic get shot. People who stay cool stay alive. When the two protagonists—who have managed to stay cool despite flying bullets and an overdose—finally *do* lose their cool, they reach outside their own movie for help. Harvey Keitel, Central Casting's quintessential goodfella, attending a dinner party that looks as if it is occurring at a different time of day and even in a different movie altogether, agrees to assist the two younger mobsters in getting rid of a body. Tarantino himself is present at the clash of genres, playing a nervous househusband who wants all this resolved before his wife gets home and sees the bloody mess. It's Tarantino wearing the movie's wristwatch, desperately hoping his characters can conform to linearity long enough to get the job done. The stress imposed by linearity is the source of the comedy. Time itself is an antagonist.

By the end of the movie, it is balding actor Bruce Willis in a battle with time. In flashback, we learn through Christopher Walken—pulled right out of *The Deer Hunter*, in which he played an identical role—that Willis's prize possession is his father's watch. This heirloom, a comically blatant representation of time and lineage—has been through an incredible journey. As a POW in Vietnam, Willis's father hid the watch—which had belonged to his own father—in his anus. Before he died, he gave it to his buddy Walken for safekeeping. Walken, in turn, hid the watch in his anus for several years before returning to the United States and passing it on to Willis. Now the watch belongs to Willis, whose girlfriend accidentally leaves it behind when they are running away from a mob boss.

Willis risks his life to go back for the watch, kills the henchman waiting for him, who happens to be on the toilet reading some pulp fiction when Willis gets home, and then, completely coincidentally, runs into the mob boss. Another chase scene later, and Willis and the mob boss both find themselves prisoners of S&M, war veteran rapists. Willis struggles free and, against logic, goes back to rescue his mob boss pursuer. It's as if the rapists are from a different movie genre, so the gangsters have to stick together. After successfully rescuing the mob boss, whose own anus was being violated by men in military uniforms, Willis is pardoned for his sins and allowed to go free.

So even in the world of *Pulp Fiction*, chaos turns in on itself and a new sort of order rules. By reaching back in time for his own lineage—the family watch—Willis began a quest for continuity against all odds. That quest ironically yielded a number of chaotic coincidences, freeing him from his own mistakes of the past. Tarantino also offers us his take on how and when time and lineage got so difficult to deal with in a straightforward fashion. It was during the Vietnam War, when traditional American values faced their greatest challenge: the television image. The daily TV dinner of images with American boys dying in a war no one understood was just too much for the cultural conscience. The pictures broadcast through our Philcos began to recast our moral templates. Children rose up against their parents, students against their teachers, and a people against its government's policies.

This frightening and disillusioning adolescence led to two very different coping strategies. *Forrest Gump* and *Pulp*

Fiction both attempt to reckon with the implications of the post-Vietnam era—the relativistic haze of a world whose history and self-image is defined by media. Because the television image is so nonmaterial, ethereal, and malleable, it is extremely difficult to get one's bearings in a world made up of little else. How do we self-define, as individuals and a society, in a culture made of images?

Gump's answer is to paste them all together, even after the fact, in order to tell the story you want to hear. As long as you play stupid, that will work. Gump is ever the child of his mother, whose platitudes ring on long after her death. He remains true to his childhood programming, with his character and actions predetermined.

Pulp Fiction and the child of chaos's postmodern answer is to accept that there is no sense to the sequential progression of the images. Rather than being an adult child who wanders through life on faith, be a child-adult—a child of chaos—who surfs through life accepting the responsibility of self-determinism and the grace of nonlinear experience. Equally important, the screenager accepts the media technology generating all this confusion as a partner in forward evolution. It is not something to fear, but something to embrace. If we do learn to accept the turbulent nature of chaos, we may be surprised to find that we are not simply subjected to the random whims of an indifferent universe, but rather plugged in to a much deeper sensibility about what is ultimately holding everything together.

3

THE FALL OF MECHANISM
AND THE RISE OF ANIMISM

"If you started in the wrong way," I said in answer to
the investigator's questions, "everything that happened
would be a proof of the conspiracy against you. It
would all be self-validating. You couldn't draw a breath
without knowing it was part of the plot."

Aldous Huxley, *Doors of Perception* (1954)

As if to convince us that the digital revolution poses no threat
to the sanctity of flesh-and-blood reality, traditional computer
technology theorists like those out of MIT's Media Lab are at
pains to differentiate between what they call bits of data of
the informational world and atoms of the physical world.
There are bits, and there are atoms, and never the twain shall
meet. In an effort as misguided as the original attempt to separate religion and science or mind and body, the "wired" academics are revealing how they are at cross-purposes with
themselves: How are they to spread digital culture while keep-

ing control of it? How does one disseminate an intrinsically empowering technology without losing status as its exclusive purveyor? In other words, how does one maintain the institutional sanctity of an elitist computer lab once all of society has itself become a computer laboratory? By keeping technology mysterious. Unreal. Secret and apart from our daily experience.

The modern-day secret societies, cabals, and other elite groups maintain their hold over seemingly magical technologies by making them appear beyond the grasp of unindoctrinated persons—outside commonsense reality. The artificial and utterly false distinction between matter and information—body and mind—amounts to a division of realms. Regular people can only hope to understand the physical reality, and must turn to priests and academics for some interpretation of the other ones. To become one of those chosen few who gain access to the privileged information, one must make a deep, personal investment in one of the secretive institutions. This can mean years as an assistant professor, associate editor, or junior board member. By the time one passes through the hazing, one has a stake in the institution picked. This is the same way a cult indoctrinates new members: By asking them to perform progressively outrageous acts—against their common sense. The more extreme the act of commitment, be it prostration before a guru or donation of money, the more the initiate needs to believe his actions are dedicated to a worthy cause.

The cultlike process by which our scientific and academic institutions indoctrinate new members serves to preserve the values and status of the cabals. It prevents evolution. But we, the general public, are not merely the unwitting victims of an academic conspiracy. We like it this way. It's nice to believe that someone knows what's going on. We appoint our own pundits as willingly as they accept the mantle.

The keepers of our higher, secret truths protect us from

the devastating implications of our newfound technologies: namely, that it is becoming difficult for us to distinguish between the physical and the digital, the scientific and the spiritual, the mundane and the magical. This is not because there is some set of differences that we cannot perceive. It is because they are all indistinguishably and intimately related. Even information on a computer disk has a physical component; certain sectors are magnetized and others are not. Information, whether indicated by the positions of beads on an abacus, the charge of a byte of RAM, or a codon of DNA, has definite physicality. Meanwhile, the real world we touch and observe every day is getting less definite all the time. We hardly know if atoms are matter or energy, or if any physical particles are real at all. We can't honestly say that information isn't real stuff when we don't even know what real stuff is.

Our answer to this dilemma so far, as aided by our anxious punditry, has been to deny the connections between the real and the digital, or the physical and the magical. Biology may evolve, and technologies including magic may evolve, but they are to be considered mutually exclusive provinces. In stark contrast to this supposition, however, kids' culture stands as a delightfully mixed-up common ground for all these digital, magical, and biological sorts of development.

When we look carefully at the reaction of younger cyber-denizens to their Sega environs, we find that they make no distinction between information and matter, mechanics and thought, work and play, or even religion and commerce. In fact, kids on the frontier of the digital terrain have adopted some extraordinarily magical notions about the world we live in. Far from yielding a society of coldhearted rationalists, the ethereal, out-of-body experience of mediating technologies appears to have spawned a generation of pagan spiritualists whose dedication to technology is only matched by their enthusiasm for elemental truth and a neoprimitive, magical worldview. To a screenager, these are not opposing life strategies but coordinated agents of change.

Real Stuff

Parents' most common fear as they watch their children sink deeper into the artificial worlds of Doom videogame dungeons and America Online chat rooms is that the kids will lose touch with the physical world. They won't want to play outside anymore, and when they grow up will no longer cherish live interaction with other human beings. Much to the parents' astonishment, however, the more technologically mediated a kid's lifestyle, the more he longs for contact with physical reality, and the more he values real objects for their authenticity.

Whenever a new technology arrives, we learn the true nature of its predecessor. When videocassettes became popular in the 1970s, the film industry worried that we were about to witness the demise of the movie house. In an effort to compete with the variety of films available at the rental store, movie houses broke themselves up into four-, six-, and ten-plexes. We were asked to crowd into tiny rooms to watch the movie of our choice on screens not much bigger than our own TV sets. Eventually, though, the movie houses realized what movies have that home videos don't: size, sound, and spectacle. They revamped their cinemas, elaborately redecorating their lobbies and implementing 70mm widescreen, Dolby, and THX. The movies draw in more money now than ever, because they have capitalized on what is unique about the theatrical film experience. Watching videotapes at home made us value the magic of full-screen, first-run cinema.

The emergence of digitally stored and electronically distributed data has made old-fashioned printed media much more valuable to us. The fact that we can read text on our computer screens changed our perception of objects like

110

books and magazines. More than mere holders of data, printed media are tangible objects with texture and weight, providing the sense of a physical connection to the writer and subject. Special first editions and signed copies are becoming more common and more collected. Publishers are spending more on art direction and printing. Bookstores, too, are capitalizing on the advantages of book browsing over net-cruising by investing in fancy wood shelves, coffee bars, social areas, readings series, and elaborate displays—all to highlight the flesh-and-blood reality of the book world. Book sales are at an all-time high because books' physical component and totemic value have only become apparent to us since the advent of matterless data. The weird, unknown, and almost supernatural quality of books comes from somewhere between the information or stories they relate and their physical connection to the teller and the teller's tales—the sense that the book as an object came from the writer. We can take a book to bed with us and, by inference, its writer. The World Wide Web may allow us to click on an icon and magically reveal pages of text, but it only makes us more aware of how books themselves are magical objects, with an unquantifiable set of resonances to them.

Likewise, the further kids' heros and cultural icons remove themselves into the media strata, the more kids value the magic of the original, genuine, physical articles associated with them. For example, the bigger a celebrity, say Madonna, becomes, the more valuable is her genuine autograph, or the first pressing of one of her records. As she becomes more available in the mediaspace as music and image—information—then the real objects representing her take on a totemic value. Counterfeits will not do. The ease with which digital information can be copied, transferred, and electronically mailed and broadcast changes the inherent value of the physical.

Although many American marketers have begun to complain about the emergence of bootleg merchandise from China, most kids strive to purchase genuine, authorized versions of their toys and collectibles and welcome the manufacturer's efforts to legitimize the authenticity of their products. In the 1992 basketball championships, for example, winning players were awarded baseball caps with their team's insignia. Each player wore his new cap complete with a large, 3 x 3-inch NBA tag hanging off it. A small, irreproducible holographic image on the label guaranteed the cap's authenticity as a genuine NBA product. Over the next few weeks, kids purchased the new official NBA caps and wore them without removing the tags. This was the new fashion. A *real* NBA hat means more than a fake one, even if they are identical. The kids could as easily be wearing the tag around their necks with no cap at all.

This is why the cartoon world of animé, ethereal in both its subject matter and execution, nonetheless yields some of the highest-priced collectibles in the toy industry. Although images and short clips of animé videos are readily available for downloading off the Internet, kids will pay hundreds of dollars for original tapes, toys, models, and posters of their favorite characters. They have learned to recognize the identifying trademarks of legitimate versions of their toys and shun cheap knockoffs, such as models recast from originals. This shouldn't be so surprising. Shows like the *Power Rangers* are based on the idea that objects have intrinsic, magical value. They can be inhabited by human consciousness. If kids are going to pretend they can inhabit a gundam robot toy, it had better be genuine. The toys are supposed to be directly linked to the evolution of life and technology on our planet; however discontinuous the media portraying them, these toys must have a direct, physical link to the continuity of their development.

Pogs—those bottle-cap collectibles—are probably the most extreme example of kids giving totemic value to physicalized

versions of media iconography. These are coins of the realm—
the pop cultural realm—and their value is based solely on the
perceived value of the media icon encapsulated on the pog.
Unlike coinage, pogs do not carry pictures of rulers or
national symbols, which help legitimize traditional currency.
No, they carry the images of cartoon characters, sports heros,
clothing labels, and television networks. A child's arsenal of
pogs is a working model of his cultural values. It is the conceptual equivalent of his genes, which he then mixes and
matches with other children's when they trade or battle for
pog possession by shooting them. (One stacks his pogs in a
pile, and the other player knocks them over with a "slammer,"
a larger, special pog. Whichever pogs flip over are won "for
keeps" by the tosser.)

Pog playing has become so popular that it has been banned
in many public schools across America. The passion with
which kids collect and trade their pogs derives from the resonance of their iconography, and the combined, magnified resonance of an entire collection. Like a bag of runic stones or a
deck of tarot cards used for divination, the pogs have both
physical and representational dimensions. When these two
qualities meet in a single object, it gains totemic value.
Casting pogs, gaining another player's icons, or losing icons
from one's self-selected collection is an almost genetic-style
exchange. It is an intimate interaction, not in that fluids are
mixed, but in the way a kid's collection of totems is forever
altered. As in the tarot or I Ching, where cards or coins are
shuffled or cast randomly, there is also an element of chance
at play. No one knows which pogs will turn faceup or face-
down.

This is why the pog game satisfies the more mystical aspi-
rations of its otherwise digitally oriented participants. It represents a surrender, of sorts, to natural, elemental forces as
translated through resonant iconography. The icons—the
informational aspect of the pog—are intentionally purchased

and collected; the pogs themselves—their physical presence—
are subject to the game, the wind, the surface of street, and
the force of a competing player's slammer. What we will see
repeatedly illustrated in the collections of objects and ritualized actions of the children of chaos is a marriage of technology and magic—the informational and material.

This relationship is not new, even if its expressions are.
Tarot cards and I Ching coins can be understood as technologies with magical implications. The questioner randomizes
his cards or coins, surrendering to chaos, but the cards,
spreads, or I Ching hexagrams themselves are highly ordered
systems of archetypes, passed down through generations of
practitioners. Nearly every divination tool combines an element of pure chance—a birth date, the numerical value of letters in one's name, the selection of a rock or shell—with a
highly mathematical interpretative technique like astrology,
numerology, the *Taoist Book of Changes*, or 78-card tarot
arcana. Many of these divination techniques began as calen-
drical systems. The I Ching, for example, is based on a lunar
calendar, which many still claim is more accurate than our
own Gregorian calendar and has become the basis for numer-
ous spiritual systems. Magic potions and crystals are based on
the tools of ancient alchemists who, although superstitious
and mystical, were also scientists out to determine the particular nature of the material dimension. What we today call
magic is all based on what was at one time cutting-edge technology. Meanwhile, many of our technological advances—like
radio crystals, fractals, and fuzzy logic—are based on ancient
precognition of molecular structure, self-similarity, and the
organic basis for intuition. The world we live in today is filled
with objects, technologies, and magics at various stages of
evolution.

The movement from the physical to the ethereal to the
material-magical follows a three-stage process. First, there is
the thing itself—let's say, gold. It has an objective value, and it

served as transactional currency for centuries.* Eventually, gold was replaced by paper money representing gold or another precious metal. Gold and silver certificates were, originally, redeemable for a specified amount of metal. The government had this metal sitting in a vault somewhere, and the leader's portrait or his treasurer's signature on the face of the note attested to this fact. Money, in this second, metaphorical stage of its development, became much more ethereal. The paper stood for something else: metal. Its value was based in our trust of our leaders and their repositories. When the dollar was taken off the silver standard, it reached its third, most magical level: the Federal Reserve note. As the bill no longer represented anything tangible other than "legal tender," the Treasury began to print a fascinating motto (already in use on coinage) on its back: *In God We Trust*. The bill came to represent nothing more than information. There was no longer any fixed amount of gold to back it up. Now the dollar itself was to generate its own, intrinsic value, not representing but rather performing, or recapitulating, the function of gold. In order for it to do so, though, the spiritual forces must be invoked. The Federal Reserve note itself is invested with totemic value. Even when stored magnetically as debit or credit, a dollar is linked to God, not gold. We take it on faith.

The kind of God our treasuries count on to protect the sanctity of our currency is very different from the Gods emerging out of pagan and kids' culture. For God to work as the guarantor of our currency, there had better be just one. Monotheism is the best prop for any simple hierarchy, because it puts a neat dot at the top of the pyramid (another great image on the back of a dollar bill, actually). If a government or church is going to let people start assigning magical value to objects, it had better be very careful to control those

---

*Actually, gold had no truly intrinsic value at the time other than its scarcity. The true literal stage of currency would have to be the barter system, where objects like candles, wood, or salt were traded for other goods or services.

objects. A Catholic can only accept Mass (wafers and wine that are believed to transubstantiate into the body and blood of Christ) if he is in good standing with the Church. People just aren't allowed to do that sort of thing on their own.

But as commodities and information-based totems are marketed successfully to our kids, the dollar bill is no longer the only faith game in town. When people collected coins, gems, or even pork bellies and Cadillacs, it posed no threat to ordered monotheism. These things all have intrinsic value as physical objects, even if that value is inflated by sentiment or whimsy. When kids begin to value objects solely for their connection to forces other than the Treasury or God, the government and Church lose control over the direction and allegiances of culture.

In an amazing display of self-awareness, the themes and subject matter of the more spiritual films and games enjoyed by the children of chaos directly address the conflict between the traditional and chaotic approaches toward magic. On the side of "order" are the governmental and religious entities who would control, codify, and secretize the magical strains of cultural experience; fostering chaos are the free-market capitalists and pagan spiritualists, who promote a more evolutionary relationship to magic, value, and nature.

Conjuring *vs.* Cajoling

Most adults hate Barney. We have trouble understanding how this saccharine, costumed purple dinosaur could so captivate our children. Except for the most cynical of us (like college students who burn giant effigies of the character in "Barney hate rallies"), we tolerate Barney because he teaches fairly innocuous but positive values along with the alphabet and a little Spanish. Our education specialists and child psychologists tell us that Barney is popular because he expresses unconditional love and support for his child friends, and teaches social skills and nonviolence. But that doesn't account entirely for Barney's rise to cult status and 15 million-plus viewers per week.

Kids are fascinated by Barney because he is conjured. In the first moments of each show—moments most parents probably ignore because they look like opening credits—Barney is summoned into being. He is a small, stuffed dinosaur sitting in a tire swing. The children perform an incantation over the doll with their song, and Barney grows, like a rubber-suited Japanese monster, into a human-sized dinosaur who plays with them and understands their problems. At the show's end, Barney says good-bye and reverts to his inanimate form. A twinkle in his eye is the only hint that the plush toy might be more than he appears. The marketing ploy approaches genius. Every authentic Barney doll could, theoretically, be conjured into a real Barney. The product has the magic within it.

The marketers, or at the very least the market forces driving television content and toy manufacture, yielded some quasi-pagan programming content. In an effort to make toy Barney real and media-character Barney tangible, the show's creators relied on magic. Capitalizing on existing trends even further, they took the already popular dinosaur motif—our kids easiest cultural reference to evolution, it turns out—and modified its surface features to fit the PBS agenda. But the show's producers understood what they were doing. As Barney creator Sheryl Leach explains, "We don't really think about what the parents are going to like and dislike. I think that his appeal is really somewhat magical. It's really about empowering the child. . . . Barney is a dinosaur, so it's been very natural for him to evolve."

Although they praise the show, Barney's adult fans misunderstand him completely—especially Christian social action groups, who often credit Barney with instilling "family values."

Barney does not promote family values in the traditional sense. No one on the show is related to anyone else. There is no parent or leader. The world of Barney is spontaneous, free-wheeling anarchy. It just so happens that this looks much more friendly than people would imagine. Barney teaches community values. A group of equals is brought together through a magical ritual and the common purpose of learning and having fun. In a sense, the bond experienced by Barney's comrades is extraordinarily antitraditional-Christian, and certainly anti-Creationist in its implications. The kids have reverted to medieval practices, dancing with dragons and looking to their community and their magical skills rather than to their leaders for guidance. The alphabet, Spanish, and counting are just a few more technologies for them to take with them as they evolve into their empowered future. And all this, brought to them by the marketplace.

A few clicks of the cable remote will bring you to more advanced versions of these, frankly, subversive values—this time woven into the stories of Hong Kong cinema, a magic-filled genre that is growing increasingly popular among American teens and twentysomethings, as well as influencing mainstream movies and commercial television. Video store shelves are bursting with titles and the World Wide Web is fraught with sites dedicated to chronicling the hundreds of films available. On the surface, these are simply martial arts action films. Quick cuts, daring stunts, sexy stars, and special effects highlight early-nineteenth-century tales of warring lords and their loyal students. Every lord teaches a different martial arts style, each with its own techniques or magical elements. Conflicts develop as the strength of one magic or technique is tested against another, and plots usually hinge on the new combination of techniques with one another, or the discovery that a magic power might be able to be used in a new way.

Like the constant development of gundam prototypes and

new-type children in animé films, Hong Kong cinema depicts a world where the fighting techniques of the ancient martial arts schools develop along with and in contrast to the spiritual and magical skills of young, heroic students. Strangely appropriate for the modern cultural climate, the plots often pit evil lords, using highly technical weapons to support corrupt government officials, against gentler teachers and fresher students, who have more of a connection to nature and spirituality. A cursory analysis of almost any of the major martial arts films from the 1980s or 1990s will reveal an anti-elitist, pagan, populist, and almost libertarian attitude toward the dual evolution of magic and technology.

One of the most popular Hong Kong film series of all time, released by leading producer Tsui Hark, began with a movie called *Swordsman*. The story opens as a sacred scroll detailing an invincible martial arts technique is stolen from a government-sponsored library. The evil lord in charge, a government operative, immediately initiates a cover-up and blames an innocent rival for the crime. The hero of the film, a young swordsman, must evaluate each of the forces in the struggle, even his loyalty to his own teacher, in order to determine which alliance to fight for. The government side uses hired mercenaries and twists the law to its own advantage. Their object is to maintain the centralized authority of the incumbent regime and keep the knowledge of the sacred scroll to themselves. Their strength comes from established laws, as well as their connection through lineage to powerful secret fighting stances and pyrotechnics.

The rebels are an alliance of disparate clans. Because they are so decentralized, they must fight whenever they meet in order to prove their identities to one another. Martial arts styles take so many years to develop that only a legitimate student of a master's house could fight with that particular skill. The rebels' fighting techniques are based in nature—the wind, mountains, elements, or even snakes. Unlike their enemy, who

119

is dedicated to preserving his power for generations, they enjoy the temporal, drink wine, and sing about "the here and now." While the government works to monopolize commerce and keep women in a subservient social role, one of the clans, run by beautiful female warriors, operates a successful smuggling business.

The allies of the government use established laws and social codes to their advantage. An evil teacher, in an attempt to force his student to reveal a secret and break a solemn oath with an old friend, cites the school's rules to which the student has also vowed allegiance. According to one rule, the boy should honor his teacher's wishes; according to another, no one deserves more loyalty than the teacher. Luckily, the wily student knows the rules better than the teacher, and outsmarts the old man so handily that the teacher finally claims, "I was only testing you!" A government official goes so far using established laws and social codes against adversaries that he deputizes one of his enemies so that he cannot fight against him. Meanwhile, the rebel allies depend on the forces of nature. In the final battle, a clan of women smugglers calls on a swarm of poisonous bees to sting the government agent to death. Although, officially speaking, the victorious alliance of renegade clans have behaved in a manner traitorous to the state, the audience understands that they have answered to a higher authority and are working in harmony with a greater force.

"It's not a matter of rank. How can you lie and betray so many people?" the hero asks the villain, forcing him to account for the huge number of people he has murdered in order to increase his personal power. "*He* is the true traitor," the young swordsman realizes before the closing credits.

Although its mode of expression may seem simplistic, the movie and most of the martial arts genre stand as a clear condemnation of the "guardian morality." The victorious young people trusted one another enough to form real alliances and depended on the magic of nature and strength of community rather than the secrets of technology and enforced lineage.

These themes recently crossed the Pacific Ocean as a cult film called *The Crow*, starring the Chinese-American son of Bruce Lee, the late Brandon Lee. The script for this dark action film came from a popular comic book novel—the first medium to promote the values and aesthetics of Hong Kong cinema stateside. Although the story is very different in plot and texture from a movie like *The Swordsman*, *The Crow* neatly translates the Hong Kong cinema's magical agenda to the American experience and its youth culture.

The world of *The Crow* is called "punk-goth"—a term coined by comic book artists combining the punk and neogothic styles to denote a romanticized yet grimy and anarchistic interpretation of present-day urban America. Think *Edward Scissorhands*, *Blade Runner*, or *Interview with a Vampire*. The story is about a young rock singer, Eric Draven, who comes back from the dead to avenge the violent murder of his fiancée. He pushes himself out of his grave, walks to the remains of his home, puts on white makeup and a goth leather outfit, and goes on a justified killing spree. He has returned, however, with a number of magical abilities. He is aided by a crow—his link to the living world—who allows him to see things remotely, in "bird's-eye" view. When Draven is shot, he doesn't get hurt, and when he touches any object or person that had a connection to something in his life, he can reexperience that moment of connection. He is able to tell the difference between his fiancée's engagement ring and many others, for example, because he can perceive the resonance still within the genuine object, connecting it to his own past. He can hold photographs and reexperience the moments when they were taken. His connection to these objects, according to the story, is the true love he felt for his fiancée.

His fiancée, we learn through these magical flashbacks, was raped and murdered by a gang of thugs working for a corrupt landlord with ties to high city officials—an efficient translation of the Hong Kong hierarchy to the modern urban milieu. Anti–free market to the core, the villains have murdered Draven's fiancée for uncovering their plot to fix rent prices in the neighborhood. Just like the warlords of *The Swordsman*, the evil gang leader has a few magic powers of his own—powers that he gets through drugs, sex with his own sister, and the murder of young girls and consumption of their eyeballs. (*The Crow* was a comic book, after all.) Although the villain and his sister discover that by killing Draven's crow they can strip him of his invincibility, it's not before Draven exacts his revenge. When he finally kills the villain, it is by reversing his own ability to see the past by touching people and objects. He has, over the course of the film, collected many images of his fiancée's painful death. This has functioned almost as a weakness, incapacitating him each time he relives her agony. At the end of the film, when he is about to be killed by the villain, he touches the man and transfers the overwhelming psychic pain, killing him in the process. (The crow, meanwhile, has befittingly pecked out the eyes of the villain's witchy sister.)

As a movie plot, *The Crow* may leave something to be desired, but the film's themes and its narrative structure go to the heart of kid culture's fascination with magic, gothic romance, and relationship of true love to the supernatural. The film is narrated by a child who understands that the natural force of love drove Draven's rage and provided him with his powers. The story is filled with totems and objects that generate real energy, and concerns a conflict between young lovers and well-connected but corrupt landowners. Most important for an understanding of the new Western magical sensibility, the tone of the story is gothic—or goth-romantic. The style of clothes, makeup, and set decorations is that of a goth club, where the romantic aesthetic of England in the eighteenth century is reinterpreted with the leather, puffy white shirts, and jewelry of the 1990s. Draven quotes the poetry of Wordsworth and Shelley, and nearly the entire movie occurs at night, in the rain. Further grounding themselves in the culture of chaos, Draven draws a giant graffiti crow after each of his murders, and the little girl narrator of the story skillfully scoots around town on a skateboard. Ironically and rather morbidly satisfying the young audience's passion for self-similar angst, Brandon Lee was accidentally shot and killed during the making of the film, mirroring his character's and his father's untimely deaths.

The increasing popularity of cinema of this sort exposes our screenagers' quest for genuineness, spirituality, and love in the face of cold urban and cyber postmodernity. Their presumptions about life follow naturally and almost predictably from this quest. Physical objects can have spiritual essence. Magic is superior when it is connected to love and elemental forces. The monopolization, centralization, and control of human interaction or land and property is to be resisted. Techniques and technologies that are separated from nature or made into secrets become destructive forces. The ability to experience empowerment in the present moment, spontaneously, is worth much more than power derived from connection to a lineage or social hierarchy.

There is another way for a magical sensibility to arise out of technological advancement, and we are witnessing that in the overculture of adults. The mystification and elitism of technology, and the hoarding of its powers by self-appointed experts, be they the faculty of MIT or the editorial staff of *Wired* magazine and the high-powered corporate CEOs on its covers, creates a superstitious public. This may even be their purpose. As the *New York Times Magazine* recognized in an otherwise admiring piece, "The message is clear: Without *Wired*, you'll drown. The magazine makes the future look like a terrifying, disorienting place. This is no accident." Or, as *Wired* writer and *Media Lab* author Stewart Brand sincerely explained for an interview in the *Los Angeles Times*, "I think elites basically drive civilization." Technology for the rest of us, then, is granted by the grace of these elites. Similarly, the assumption of spiritual superiority by priests, mystics, or the digital elite makes enlightenment or salvation feel like states that are bestowed not achieved, and certainly not a birthright. All kinds of technology, be they ancient magical systems or modern computer networks, are made to seem remote and unattainable to the average person. As a result, our attitudes toward both verge on the paranoid and conspiratorial. But we'll get to the New Age movement a little later.

A youthful, participatory, streetwise, and cyberpunk attitude toward technology yields a natural inquisitiveness toward other initially elusive but absolutely embraceable aspects of life. Maybe this is why the kids who demonstrate the greatest willingness to experiment with new technologies are the same people who experiment with some of the oldest ones.

Bloodlust

Today's movies, no matter how elaborately they expose the magical inclinations of our kids, are only a surface reflection of a much deeper, instinctual drive to incorporate a spiritual sensibility into the modern experience. Because this sensibility is a departure from traditional religious lineage and handed-down truths, its expression looks very different from what we have come to regard as religious faith. For a spiritual practice to work in the culture of chaos, it must indeed be a practice and not a passive absorption of dogma or morality. It just so happens that the only place kids are allowed to practice their faiths without alarming adults is in their play.

Games, ritual, and theater that to past generations granted only vicarious thrills have been transformed by our current generation into participatory forms of entertainment. Even the word *entertainment* (literally to "hold within" or "hold among") refers only to a captive-style experience, where plots and characters maintain an audience's interest so that it can be held within a given structure or thought system long enough to consume the moral. The audience or congregation is gathered to receive and not partake. For kids today, spirituality or even just simple connection to elemental reality (which may be the very same thing) can only be experienced through participatory activities. It's as if they understand that the more secretive or handed-down a spiritual lesson, the more paranoid and harebrained its ultimate expression. The newly popular game-playing activities of youth culture reinvent not only the subject matter of the spiritual quest, but its mode of conveyance.

Technology has pushed kids toward spirituality by inspiring a need to reconnect to a more romantic physicality and by provoking a willingness to participate more directly in games and rituals. When the two urges combine, we get fantasy role-playing, totemic card games, macabre live-action antics, and even some genuine bloodletting.

It all began with Dungeons and Dragons, a game originally played by nerds. Only the math geeks had enough patience to learn the complex system of character points and combat scenarios, as well as a strong enough desire to find an alternative to a social scheme where they always came in last. But the game's subject matter and playing style soon attracted many many more to the "fantasy role-playing" genre. FRPs, as they're called, allow players to create their own characters and then make up stories about them as the game is played. Players join in the game by creating a character sheet. Each player has the same number of points from which to construct the character he will play in the game. Every skill and level of proficiency costs a certain number of points, so each player must choose whether to spend his points on physical strength, magical abilities, dexterity, speed, beauty, or any other quality he feels will help him move successfully through the fantasy world. A player can even choose a handicap, like a missing arm or a drinking problem, in order to gain extra points to use elsewhere.

Together, and sometimes with the help of a referee called a "gamemaster," the players invent a story. Each character has his own set of goals within the context of the story, whether it's to find a hidden secret, overthrow a tyranny, or simply have fun. When interests conflict, combat of one sort or another ensues. The players then fight using whichever skills they gave themselves on their character sheet or may have developed over the course of the game. This part of fantasy role-playing is pretty mathematical, and participants must roll dice and then calculate the outcome of a battle based on the arithmetic value of their character's strength or technique. The majority of the game's action, though, works like an improvisational acting exercise, with the players seated around a room or at a table, performing ad-lib dialogue with one another and describing their own actions. Some groups use little figurines to represent the characters, which are especially helpful in choreographing and visualizing multiplayer combat scenes, while other groups simply imagine how the scenes and characters look.

Because of its free-form nature, the game functions almost like a collective lucid dream. Although the players strive to behave "true to character" and might have their actions censored by the gamemaster if they don't, the story and its characters develop on their own. Beginners and less imaginative players can purchase prewritten scenarios at a game store, but even these allow the players a lot of latitude. The joy comes from generating a story together and watching the group imagination unfold. Also like a dream, the milieu of Dungeons and Dragons allows for wild fantasy and highly symbolic interactions. A mix of Arthurian and Tolkienesque mythology, the magical landscape of goblins, wizards, brutes, dragons, and damsels is a mutable, magical world where almost anything imaginable can take place.

Most committed players are remarkably well read, especially in authors like C. S. Lewis and J. R. R. Tolkien, and well versed in Jungian archetypes, medieval lore, alchemist's jargon, and other highly totemic systems of thought. Adventures include pagan rituals, magic spells, interdimensional travel, mysterious keys, objects with totemic powers, companion animals, and individual vision quests within the overall group vision quest; they move like guided visualizations, from choice to choice, where each gate, key, chair, rodent, and cloud has the resonance and possibility of a dream symbol. No one, except maybe the gamemaster, inhibits the imaginative journey, which naturally progresses toward chaotic group fantasy and wish fulfillment. This worries parents, who see their children as retreating further into a world of paganism, far removed from consensus reality. Worse, they witness their kids becoming increasingly confident of themselves the further out they get.

Fantasy role-playing—like traditional religion and formal psychotherapy—serves as both a spiritual practice and transformational tool. The difference between the games and standard religion is that FRPs allow kids to explore their spiritual outlooks purposefully and without repercussion. The kids playing these quasi-religious games every weekend understand that the imagery is, indeed, imaginary. The gods and goddesses in their games are of their own invention and can be created and destroyed. Obeisance is not mandatory, because damnation in the world of the game has no effect on the player's real life. Well, at least not the same effect. As play, the game offers kids the opportunity to very consciously test out new life strategies. Someone who solves problems in real life by physically fighting may choose to play a character who battles using his wits. Someone who has developed a mechanistic attitude toward life may use the game to explore more magical worldviews. Unlike psychotherapy, though, the games don't simply curb neuroses so that the player can more successfully adapt to consensus culture. The fantasy world around the players is as malleable as their individual strategies, and it changes as the players adopt different tactics. If players decide to solve problems through violence, the world becomes violent; if they choose to use magic or technology, then those skills become more valuable over the course of the game.

Since FRPs promote an evolutionary model of culture, the games themselves soon expanded to include many kinds of worlds other than ancient elfin ones. Nearly every popular genre from the youth zeitgeist has been packaged as a game kit, from cyberpunk and animé to tech wars and the sex industry. The books describing these play worlds are so realistic that a cyberpunk game about computer hacking was once mistaken by the Secret Service for a real instruction set on how to commit computer crimes. The most advanced games, usually developed by individual FRP player groups, combine the aesthetics and technologies of many different worlds. By allowing players to travel through time or interdimensionally, a single game can include ancient wizards, punk rockers, *Star Trek* characters, and alien life-forms, all conversing and combating using weapons and tactics from different eras. Gamemasters sometimes have fun by declaring that, within a particular dimension or on a specific planet, nothing more technologically advanced than a crossbow will function. Characters relying on high-tech tools will then have to fight with their fists.

Bringing so many apparently different genres into the same malleable and evolutionary game world tends to make them interchangeable in the minds of the players. The magic spells of a wizard are no more spiritual than the programming skills of a computer hacker, and no less technological. They are simply different, and draw their power from different sources. A good player can skillfully move through a fantasy world with almost any character sheet. The game teaches how to use whatever resources may be available in any environment, even a changing one. The magical, spiritual, or cyber technologies can all be used, and in any combination, as long as a player has had the foresight and the gumption to include them on his character sheet.

FRPs based on cyber scenarios and comic book worlds soon attracted fans and participants of those media into the action. Hybridizing the arithmetics of standard role-playing games with the collectibility of pogs, a mathematician named Richard Garfield developed a card game called "Magic: The Gathering" in 1993 to appeal to these burgeoning new markets, and sold over 10 million cards in the first six weeks. Each player takes the role of a wizard, but instead of creating a character sheet, he collects cards that become part of his playing deck. Like a collection of pogs, the selected 60 cards, of the hundreds available, make up the wizard's arsenal of spells, creatures, and enchantments. To get more and better cards, the player needs to go to the store and buy them randomly from sealed packs. The cards fall into three main categories of availability: common, uncommon, and rare, with the most powerful cards being generally harder to get.

The wizards play against each other by summoning creatures that attack their opponent, casting spells to disable their enemies, and employing "artifacts," which are magical objects with a variety of malicious and benevolent functions. Although the world of the game is pure magic and wizardry, its play is highly mathematical and requires the development of a personal strategy. Some players concentrate on developing a deck with many spells and artifacts but few creatures, while others build "combat decks" almost entirely dedicated to summoning strong monsters and attacking the opponent. A player's game strategy even extends to the way he purchases his cards. Some kids buy as many cards as possible and create dense and powerful decks. Others look down on this substitution of money for brains and prefer to create decks composed exclusively of "common" cards. A player *is* his deck. Kids carry their decks with them to school, and constantly "tune" them by replacing certain cards with others. The marketing, collecting, and playing of Magic are all interdependent aspects of the game, which is one of the reasons the game has sold so well.

Game store owners have dedicated large sections of their shops to the Magic players, who purchase and trade cards, and even play games with one another right in the store. This marketing strategy has led to the cultivation of a very large nationwide network of players. Where FRPs were generally limited to groups of kids from the same town or school, Magic, and the many other card games that have developed since, has brought together many individuals and groups who might never have come in contact. Kids play with strangers in the safety of the store environment, and trade cards and strategies with others at comic conventions and over the Internet. I've seen Hasidic boys playing with African-American girls, and twelve-year-old nerds playing against twentysomething motorcycle gang members. Most players who have moved from FRPs to card games say that the latter offer a much more fluid social experience compared to the buddies-in-the-basement quality of Dungeons and Dragons.

The high visibility of the game led a small group of adults in Bedford, New York, to form an anti-Magic coalition called Concerned Parents: Citizens and Professionals Against the Seduction of Children, claiming that the game spread Satanism. At a public town meeting, the reactionary group's founder said that the game employs "black magic and mind control. . . . It uses incantations to the devil, contact from below." Among the games' many defenders, a seventh-grader rose to explain articulately that "there are 2500 individual Magic cards. It would be hard to find 20 that were objectionable. There are no chants, no sacrificing, no death and—definitely—no contact from below."*

The real threat that a game like Magic poses is to our debilitating cultural presumptions. It combines an ancient spiritual aesthetic, a highly mathematical combat sequence, a free market collecting/trading strategy, and a broad-based, non-prejudicial social scheme. Even so, it represents only an intermediate stage in the evolution of magical gaming rituals.

Gothic, Bloodsucking Geeks

The Camarilla is a game. Well, sort of a game. The most participatory expression of the 1990s vampire craze, bookended by Ann Rice on one side and an AIDS crisis on the other, the game is one of the fastest-growing alternative social scenes since the rave, drawing participants from groups as diverse as Deadheads, computer nerds, theater enthusiasts, high school students, trust fund kids, fashion plates, and gay cliques. Based on a fantasy role-playing game called "Vampire: The Masquerade," these nighttime gatherings, occurring in dozens of cities across America, are part goth club, part geek fest, part vampire sanctuary, part community theater, and, perhaps more than any of these, a ritualized, eroticized church social.

In each town where the game is played, upwards of fifty "vampires" gather at a preordained public atrium to conduct a gathering of the Camarilla, an alliance of vampire clans established to promote harmony, allot "feeding territories," and settle disputes. As in any fantasy role-playing game each player has created a character with his or her own desires, peculiarities, strengths, and weaknesses. All of the characters perform roles in the overall "world" of the game, in this case

---

*Bedford Residents Defend Schools From a Satan Hunt," *New York Times*, Wednesday, November 22, 1995.

the competing factions of vampires in a punk-gothic 1995. It's a demonic but tongue-in-cheek atmosphere of modern technology, precise Shakespearean language, stiff Arthurian formality, clever Shavian puns, and blood-worshipping pagan morality. Older vampires, for example, complain that their greater powers mean less in modern times, when a sawed-off shotgun can give a younger progeny an unfair advantage.

The world of the Camarilla appears self-consciously designed as an allegory to the modern youth experience. As the rules, available as a trade book, *Vampire: The Masquerade*, explain: "The horror of Vampire is the legacy of being half a beast, trapped in a world of no absolutes, where morality is chosen, not ordained." The game encourages developing a strategy of tolerance and cooperation: "In order to achieve even a partial victory, the characters must usually become friends. The world of darkness is so dangerous that trustworthy allies are essential. It is an evolutionary adaptation." To prepare them for alliances based in flesh and blood rather than hollow ideologies, players are advised to dispense with blind faith but respect the ties of blood: "Many modern philosophers argue that the present age has caused an apathy and disinterest in ties of faith, nation, and blood. Certainly for the Kindred, the concerns of faith and nation are as dead as ever, but none care more dearly for the ties of blood."*

The vampire world is a distillation of the most apocalyptic features of our own. Once turned into a vampire, a character's only choice is to descend slowly into madness. Insanity and death are certainties. Playing the game is simply biding time until then, as enjoyably as possible. There is no final object—not even survival—other than to increase one's sense of personal power and to stave off the inevitable decline. To this end, the vampires develop community, gangs based on blood ties (who got bitten by whom), and highly magical, totem-

*Mark Rein-Hagen, *Vampire: The Masquerade*, rule book (Whitewolf Games: 1992).

based interactions. The game is really just a stylized version of what has begun to occur in youth culture already. In the world of the game, the vampires are unwittingly controlled by elder vampires from ancient lineages, whom they never see. Over the course of the action, the younger vampires attempt to break free of this domination by getting in touch with their youthful antiauthoritarian impulses, erotic energies, and innate magical abilities—all accessed through their own supernatural blood.

The game serves the same purpose for its young players. The Masquerade is played as a "live action" fantasy game. Instead of sitting around a table rolling dice and tabulating results, boys and girls dress up as vampires and perform for, and with, one another. There are no scripts or even "objects of the game." An immense and sprawling group improvisation, refereed by the storytellers, who help choreograph fights, decide who gets killed, and keep the plot moving along. The object is not to win or lose, but to keep the game interesting. And, of course, to keep the pretty girls playing. Live action role-playing is for kids and a gaming culture that has reached adolescence. The most dangerous of all taboos—beyond alchemy, Egyptian tombs, and cyberspace—is the flesh and blood interaction called sex, and the Camarilla appears designed to bring this last social stigma into the mix.

It's as if the entire game environment were a designated FRP world—a meta-FRP consciously developed for the half-way socialized nerd to take advantage of his or her best attributes in order to get laid. The player's real-life character sheet may already be written, but he's certainly free to create a fantasy world where his attributes gain him a maximum advantage. If you can't change yourself, change your world. Girls and boys from about sixteen to slightly over thirty have been drawn to the social ambiance of the game. While not all of them are committed to the point values and back story, everyone appreciates a social setting alternative to the mall or pick-up bars, as well as the romantic atmosphere. Above all, the game seems to function as an aphrodisiac—by design, perhaps, on the girls, but at least as effectively on the boys, turning these otherwise geeky and nervous young men into seductive, sexual provocateurs.

On the surface, the game looks something like a costume party. Small clusters of goth-dressed vampires lurk about, giggling, arguing, whispering, or simply observing one another. It's like funhouse theater with no audience, a living haunted house, a bizarre practical joke. But at its most passionate moments, it escalates into a very powerful, erotic, and spiritual ceremony. Like the Ken Russell movie *Gothic*, in which poets Shelley and Byron psych each other into increasingly heightened states of consciousness and levels of occult behavior, this, too, is a double-dog-dare-you world where holding and maintaining one's fictional character, intense energy, and air of intrigue is the name of the game. The point is to get as frightened and heightened as possible—to touch the "edge." Each player chooses to be a monster, albeit a beautiful, sexy, and well-spoken one, who lives in a dangerous dream world of magic and ritual. The game is an opportunity to touch something real, if dark, in a participatory and flesh-based fashion. It is a conjuring of madness, sexuality, and horror, within the safety of an FRP.

Maybe the well-fed suburban adolescent goth intellectuals of 1990s America *are* our culture's equivalent to the idle rich young romantics of the 1800s. But while free time and easy money may provide the opportunity for elaborate goth horror masquerades, despair provides the impetus. The passionate Romanticism on which today's goth scene is largely based was a reaction to the cold, hard realities of the Industrial Revolution. Absinthe, nature poetry, and puffy shirts were an homage to sensuality overrun by the factory and steam engine. The modern goth movement, and certainly the Vampire masquerade game, is both a product of and answer to today's equally mechanistic cyber-revolution. Many players learn of the game through computer bulletin boards, vampire MUDs (Multi-User Dimensions), or the goth home page on the World Wide Web. But most players claim what brought them to the game was a tremendous drive to reassert their flesh-and-blood natures—however temporary—in a world where that is growing increasingly difficult. They refuse to deny the reality of their somewhat awful predicament. Instead, the isolation of their digital jobs and social lives, coupled with the ever-present fear of AIDS, has led them to glorify and aestheticize the death they feel growing around and within themselves.

While younger players consciously understand only the surface level of their play, the more dedicated college-age players deconstruct each and every movement of the story as part of a greater, spiritual allegory. The game has become an all-consuming religious practice for them, and they write lengthy documents between each meeting of the game, both in their characters' voices, as journals or letters, and in their own voices, as analysts and commentators on the game's relevance to their real lives and spiritual quests. The extraordinary social safety of the game environment allows for the venting and consideration of some of life's most frightening truths.

The real question, though, is why must they choose such a quasi-satanic milieu for this spiritual reconsideration of cultural values? Players claim that the immersion in the darkness is a key component of any spiritual faith, especially now that our traditional culture appears hell-bent on apocalypse. They argue that even Jesus is represented in most Christian faiths as a well-built, nearly naked man dying on a cross. Death is imminent; the image is both sexy and horrific. The answer to this suffering, according to most traditional religions, is an acceptance of unworthiness and denial of the erotic in order to earn grace from above. In contrast, the Vampire game ritual works to make the experience of horror much more participatory for the practitioner, in the hope of pushing through rather than succumbing to the terror. Like a Hindu who stares in devotion at a painting of the sexy goddess Kali holding a man's severed head by its hair, the gamer psychs himself into a trance defined by sensuality and horror. The objective is to summon and then embody the very darkest forces and afflictions imaginable, so that their power can be channeled and ultimately appreciated as erotic.

Piercing the Flesh—Playing for Real

This ritualized, gothic-style eroticism has extended far beyond the confines of fantasy role-playing and has brought many of its participants deeper into an already thriving youth subculture that spans traditional goth music and clubs on the mild end of the spectrum and body-piercing "new primitives" on the more severe. Blood sports, sexual fetishism, bondage and discipline, industrial music, and tattoos all fall somewhere in between. What distinguishes one part of the subculture from another is the level and permanence of one's commitment to the ritual.

The "goth scene" has degenerated into a blanket term encompassing almost anything leather or antique and European. Strictly speaking, though, in its most current incarnation the term refers to a music and style trend that began in the late 1970s as a reaction to hedonist disco and the remnants of 1960s values. Unlike the punk scene, which expressed its anger outward, the goth musicians and their fans created a much more introspective genre. The despair was turned inward—not so much as rage, but as the contemplation of decay and death. Goth survived the 1980s thanks to supergroups like The Cure, and then reemerged bigger than ever in the cyber/AIDS 1990s.

Goth style is based less on the original third-century cata-comb-building Teutons than it is on the Romantics and Victorians who revived the Teutonic styles in the 1800s. In chaotic terms, goth would be considered an intermittence or recurrence of a social strain. Each time it comes back, it echoes different qualities of its previous emergences. Today's romantics are more concerned with style, mood, and imagery than they are with the overtly religious or historical aspects of goth. They wear leather, corsets, lace, whiteface, and black lipstick. Boys sometimes wear skirts, and affect effeminate or even sickly mannerisms reminiscent of a dying or overdosed young Romantic poet. Crucifixes and ankhs abound, but goth kids are at pains to explain that these sym-bols do not signify any particular religious beliefs. They are simply totems, and if they have deep meaning, then that meaning is dormant.

The main goth activity is to go to clubs and dance to the eerily reassuring music. The lyrics often speak of blood, pain, or death ("I forget how to move when my mouth is this dry / and my eyes are bursting hearts in a bloodstained sky / oh, it was sweet and wild"—"Homesick," The Cure). Still, the feel-ings are savored for their sweetness all the same. While the kids refuse to acknowledge the specific symbolism of their crucifixes, they know they are playing out a ritualized celebra-tion of martyrdom. At the clubs the kids aren't really dancing. It's more like these impeccably dressed goth kids are floating spirits in a graveyard—or each dancing alone in his own sub-urban bedroom. Death is still the pervasive ambiance, but it's been somehow aestheticized into a delicacy. Everything grue-some has been so "vogued"—self-consciously styled into par-ody—that it no longer represents a threat. The otherwise non-specific anxiety of death and even domination has been fully aestheticized and incarnated, as if to say we are all vampires, or worse, living off life yet dying all the same. Let us revel in the tragedy, and push through the despair.

Status in goth culture is dependent on one's level of commitment to the style and the martyrdom that may result. Wearing a skirt and whiteface during class in junior high school is a much more serious social commitment for a boy than dressing up for a club, and sometimes gets the kids ostracized or beaten up by their peers. Whatever it costs at school, however, is sure to be earned back at the club that weekend.

Kids intent on proving their absolute commitment to goth usually move beyond clothes and makeup into tattoos and piercing. These relatively permanent body modifications and their ritualized application are the closest activities to religious practice in this corner of the chaotic culture and have different sets of meaning for different participants. The common impulse, though, is to take charge of oneself. Self-modification, even self-mutilation, is at worst—just like listening to introspective, downer music—a controlled descent. It is a turning inward where, rather than just contemplating your navel, you stick something in there. At best, piercing and other body modifications are an initiation of conscious corporeal evolution. Designer physique.

For the "modern primitives" piercing is an initiation. Without facial piercings, you haven't demonstrated a commitment to the tribe. The more daring the piercing, the more status one can gain. This can be expressed by the size or location of the piercing, or even the pain level associated with its execution.

Most American cities today have at least a few tribes of new primitives, often street kids, who wander about parks, panhandle for money, and demonstrate as much commitment to one another as they do disenchantment with the rest of us. Modeling themselves after kids on the dole in the United Kingdom, the current crop of urban primitives live off the loose change of a system they hope will ultimately collapse. However incoherent their social theory, these kids certainly typify the way dropouts of our society's most highly ordered systems revert to the most ancient of tribal customs. Their simple gang, just like the metal in their faces, is the only constant in an otherwise unfixed and undependable world. Most of these kids felt unloved or were abused by their parents, and with nothing else to hold on to, save their self-selected tribe, they have embedded handles and footholds in their own flesh.

Others kids pierce for the erotic thrill. They claim that every new piercing creates a new erogenous zone, and that piercings on the genitals create sensations beyond description for both partners. The revival of the bondage and discipline sexual culture, as even the high school kids experimenting in it will attest, is a direct reaction to AIDS, the illegality of drugs, the unavailability of money, and the decrease in funding and interest in the arts. Alternative sexual exploration—especially if it doesn't involve fluid exchange with others—is the only mode of expression left, and the only ritual capable of inducing a "bliss response." The interaction can be as simple as a phone call from a girlfriend commanding her boyfriend to put on a certain color shirt and be at her house in fifteen minutes, or as elaborate as mutual scarification with knives, or the imprinting of a tattoo of one person's totem on his lover's genitals. Even with no partner, piercing and tattooing is a celebration of one's physicality and sexuality. To walk down the street with a stud in your nipple or penis—visible or not—is a statement of sexual self-worth. Or, at least, as more than one kid has told me, "It's the only thing we have left to be in charge of."

Piercings and tattoos are like the power spots a graffiti artist places throughout his urban terrain, except they tag the surface of the body. Each piercing holds a totemic object and serves as a pathway for that object's energy. Jewelry aside, each hole increases the surface area of the body. It turns something that was once interior flesh into an exterior and exposed face. What looks like mutilation to many adults is experienced by the participants as willful opening of the flesh. While most people are impressed by the permanence of a tattoo or piercing, these signify just the opposite to the wearers: the mutability of the human form. Like the branches of a manicured bonsai tree, one's flesh is movable, teasable, stretchable into new, intentional shapes. The skin, rather than simply being a record of injuries and pimple scars or an indication of race, becomes a canvas for tattoo art. The skin paintings are a work in progress, constantly being modified, extended, and juxtaposed with new imagery. If there is a permanence to an individual icon or image burned into the skin, it reflects only a willingness to express an idea or aesthetic officially "on the record." Filling one's skin with tattoos is to create a personal fossil record. Yes, each printed icon is permanent, but it only reflects one stage in the evolution of the individual and his layers of favorite totems.

For next century's generation of would-be cybernauts, these totems are place-holders for fully functional flesh enhancements of the future. Doctors already use implanted mechanics such as pacemakers and artificial heart valves to make up for failings of the body and its systems. Enhancement technology may not be that far off. In the past, drugs developed for treating illnesses were usurped by healthy people looking for new highs, better muscle tone, or faster recall. There's every reason to suspect that medical technology will take the same course, with implanted devices serving to enhance human functioning. Sure, breast implants are already with, or rather within, many of us, but the silicon worship inherent in such modifications is largely unrecognized. Implanting X-ray lenses, digital clocks, or microchips in the brain for extra memory or processing speed is a much more direct acceptance of technology as an augmentation of human biology.

What piercing doesn't prepare us for, books like William Gibson's do; characters accept "wetware" and "wireheading" the way we do vaccinations or a toupee. A kid who already has twelve rings in his face will have an easier time replacing one of them with an electrode or socket than someone who has never pushed metal through his skin. Even as place-holders, though, piercings still serve as spiritual totems. As these totems evolve, their iconography and sensibility is changing from goth to tech. A spirituality of self-modification and conscious evolution will move quite naturally to encompass technology as a force of positive change. In the most pagan and seemingly primitive of our kids' spiritual rituals, we find them reaching toward a coevolution with technology.

Evolutionary Abductees

Those of us who are still afraid of technology and the vengeful God whose function our machines will soon usurp, are having a much harder time coping with how magical our lives feel these days. Spiritual techniques have always been kept locked away and out of reach. The magical aspects of our technology, too, have been sequestered by government labs and ivory-towered scientists. Now that we have access to many of these practices and processes, we have begun to panic. This panic expresses itself as superstition, paranoia, and guru-addiction. The overculture is truly much more spiritually crazed than the kids.

As we slowly emerge from centuries of spiritual repression and technological secrecy we find ourselves ill-equipped to make intelligent choices for ourselves in these areas. While kids have begun exploring these disciplines on their own, most adults still consider religion and science the province of elites. In the days of Western imperialism, which are barely over, it was the government-sponsored explorers who would venture to the Far East and bring back inventions like gunpowder and paper money along with mystical practices like yoga and occultism. While the inventions were shared with the court, the mysticism spawned cabals and secret societies. Aleister Crowley and other figures arose at the center of such cults dedicated to documenting and utilizing the "magics" of exotic civilizations. This *Raiders of the Lost Ark*–style cabalistic competition went hand in hand with the development of weapons technology and population control. A government that could command the blind spiritual devotion of its minions and then arm them with superior tools of destruction could surely dominate the world.

The shrouded and competitive quality of mystical and technological research survived for centuries. In modern times, such research has peaked through its own opaque surface as CIA experiments with LSD to make better soldiers, rumors of Russian ESP laboratories, or L. Ron Hubbard's plans to transform the American psyche with Dianetics. Since World War II, in fact, the real mission of our own and competing government propagandists has been ideological warfare. Ideologies as such did not even exist before their function as cultural motivators was firmly established. An ideology is not just a collection of ideas, but a body of doctrine formed to promote a social or political program. It has an agenda. It is a justification for something else.

No wonder so many conspiracy theories have been born since the 1950s. When the Soviet *Sputnik* was successfully launched ahead of the first American satellites, U.S. newspapers printed stories reporting rumors of a Soviet plan to paint a hammer and sickle on the moon! We started to teach "new math" in high schools as the race to the Moon against our mortal enemies began. With technology married to mysticism, ideological warfare, and even a nuclear threat, it is not surprising that UFO sightings became commonplace. We were lost in a paranoid nightmare where the heavens held secrets beyond our imagination. And our level of fear was being stoked all along by McCarthyism and classroom duck-and-cover drills.

Unlike our kids, we were unprepared to deconstruct the hysteria into manageable component parts. We had no tools for reckoning with the experience—only experts, priests, and rulers to tell us what to think. Our personal spiritual wisdom had been painstakingly oversimplified during centuries of monotheism. Most of the angels and mystical references had been carefully removed from the Old Testament long ago. As pagan people were incorporated into the Church, their own gods were institutionalized as saints, each with his own little chamber in the cathedral. Their earth-based totems, like the German winter solstice tree, were co-opted as Christian symbols. Monotheism's single focus and parental style worked to maintain supply-side spirituality, but it also made its populations as devoid of independent thought as monocultural topsoil is of nutrients. We are as defenseless against confusion and susceptible to paranoia as the arid fields are to erosion.

As any meditator will tell you, one-pointedness is an important stage in the development of concentration. After a meditator learns to fix his mind on one object or thought, he slowly opens his awareness to include much more. Although we may be at this early stage right now as a culture, the (seeming) inaccessibility of the occult and technology, coupled with our willingness to accept a passive role in our own spiritual development, has twisted our urge for contact with the magical and mystical into some laughably perverse activities.

Take the New Age. This noble if naïve effort to explore ancient and indigenous spiritual traditions finds its roots in 1960s psychedelia. LSD deeply shocked the people who took it; they were totally unprepared for the mystical experience it offered. The drug brought people into a state of consciousness formerly accessible only to mystics, and then only after years of hard work. Unable to integrate the clarity, or at least the intensity, of their psychedelic experiences into waking-state reality, they turned to the East for answers. They knew they had stumbled upon entirely unexplored terrain and sought the maps of others who may have been there before.

While that in itself could have simply forced a reevaluation of Western religion and culture in the light of Eastern insights, they looked instead for replacements of what they had already: parents, teachers, and mystical leaders. Like children wandering into the dark, they reached out for an adult, any adult, with a steady hand. They were unable to reckon with the free-form quality of the psychedelic space. Even though most who went on psychedelic trips came back with the impression that reality itself is up for grabs, they were uncomfortable holding on to this perception for too long. This was a generation of kids who had grown up with leaders and institutions they could count on, even if only for negative behavior. If they were going to dispense with the old ones, they would have to find new ones. New old ones. New Age.

Scores of mystics and charlatans from the East, with exotic accents and strange demeanors, eagerly rose to the challenge. No kabbalistic rabbi or Jesuit priest, however mystically inclined, would suffice. The baby boomers were adopting new parents, and they had to be from somewhere else. Fortunately, a few of them had something valuable to share, including the teachings of Buddha, Krishna, Gurdjieff, and ancient technologies like the I Ching, tai chi chuan, astrology, and tarot.

Although these teachings and self-exploration techniques can actually empower their users, their application in New Age circles has been anything but empowering. In the New Age, you get what you pay for, and you pay to remain a student. By keeping teachers, New Agers get to keep that all-important distance between themselves and the mysticism they think they are after. It has to look weird, foreign, exotic, miraculous, and incomprehensible to be true. Marketers have no problem with this assumption. They simply translate distance to dollars. The more powerful the dogma, the higher the admission fee. In developing secular cultism, we exchange the prostration and allegiance for a game of "souling for dollars." It's just a new elite. This, combined with our paranoid, hamstrung approach to technology, has brought us into our current culturewide spiritual daze.

Each of our most perversely commercialized and superstitious beliefs—you know, the ones featured on late-night Fox programs or the wildly successful daytime talk show *Other Side*—confusedly attempts to reckon with the implications of chaotic culture. These beliefs all begin with the premise that human beings are either the victims of evil or the recipients of grace, but never the creators of their experience. Each extension of chaos or nature into one's life, then, like the extensions of computers into a technophobe's, is identified as the imposed will of some external force or being.

Angels, one of the more benign examples of external force, have descended in hordes over the New Age population. The scores of best-selling volumes about our ethereal little helpers all relate what their authors and editors insist are "true stories of heavenly visitors."* The individual stories seem less important to their archivists than the fact that these angelic interventions in human affairs are real. But not *too* real. "Angels don't submit to litmus tests, testify in court, or slide under a microscope for examination,"† explains a leading angel advocate on the back of her trade paperback collection of anecdotes. Instead, angels remain exactly where they are supposed to along the continuum between fact and fancy: in the haze. They add a little play to the steering wheel of reality. It might have been coincidence, it could even have been me, but it was probably an angel that got my baby breathing again.

Whether or not angels exist (my friends who *do* believe in them assure me that the little rascals will be amused and not insulted by my skepticism), their prominence in popular culture

*From the title of Joan Wester Anderson, *Where Angels Walk: True Stories of Heavenly Visitors* (New York: Ballantine, 1992).
†Ibid., p.6.

has a lot more to do with our need to have easy explanations for the fantastic—a personification of nature—than the cherubic effort to help us out of tight spots. Most angel stories involve a person who has found himself in a great deal of trouble—too much to imagine ever overcoming the odds and surviving. Like an alcoholic who surrenders to a higher power, the transient victim surrenders to an angel, even if that angel's earthly manifestation is a tow truck driver, a friendly "man in white," or a "touch on the shoulder" from an unseen hand. The "proof" that these influences were heavenly and not mundane always points back to something that did *not* happen. The tow truck left no tracks in the snow! The man in white never mentioned his name, and left without saying good-bye! Then there's always the standard justification, "This event exceeds the limits of coincidence."*

So angels explain situations that defy probability. Another painful side effect of the division of the realms of science and religion, and our subsequent surrender of science to laws of statistics, forces us to attribute anything that defies these laws to supernatural beings. It's a heavenly conspiracy consciously perpetrated. The division of realms disempowers us. We are either the victims of science or the victims of spirits.

Or, in worst-case scenarios, the victims of aliens. As a cultural phenomenon, the skyrocketing incidence of alien abduction begs for some thoughtful deconstruction. Americans began searching the heavens for UFOs after the nuclear holocaust of Hiroshima and Nagasaki and the start of the Cold War. The eerie potential for death from above sneaked into the thoughts of anyone who went outdoors. Man-made missiles falling from the sky and capable of total global devastation had become a reality, but the concept of human-enacted apocalypse proved too much for us to handle. Although God usually took responsibility for ordaining

*Ibid., p. 195.

catastrophic floods and plagues, he rarely exercised his will via metallic projectiles or satellites. High-tech aliens were a logical leap of faith for a society incapable of reckoning with its own destructive capabilities. The aliens within these astral vehicles quite appropriately took the form of tiny but intelligent pale humanoids with a penchant for technology: early visions of our own evolutionary future. Augmented by paranoia about government and military conspiracies, UFO-watchers concluded that the army had confiscated the real evidence of alien visitation and had hidden it in mysterious airplane hangars in the Southwest.

UFO fever resurged in the late 1970s, but now the aliens were more than just visitors; they were abductors. Since that time, hundreds of people each year report being awakened from their sleep by strange large-headed and tiny-bodied aliens. The aliens take their victims into a spaceship and then conduct painful medical examinations on them involving the insertion of long metal probes, usually in a genital orifice. Many victims conclude that their sperm or eggs have been removed for use in genetics experiments, or the breeding of new species. Others are certain that they have been implanted with tracking devices so that the aliens will be able to find them again and chart their whereabouts. John Mack, a Harvard psychiatrist whose background includes, fittingly, studying children's fear of nuclear war, has concluded that the abduction experience is real. He is careful not to say that the aliens themselves are real—only that the experiences are, and that the victims are telling the truth about what they saw and felt.

So then let's ask ourselves, in addition to the increasingly invasive quality of technology in general, what happened in the 1970s that was in the slightest bit reminiscent of strapping a person on a medical table and scratching their wombs with long metal instruments? Of course: the legalization of abortion and the proliferation of pro- and antiabortion media. The aliens themselves, with enormous black eyes and tiny, translucent bodies, look remarkably like fetuses. And the purported goal of the aliens? To confiscate their earthling victims' sperm and eggs. Again, we are watching a society confront its ability to mete out life or death. As we develop the science of gene splicing and the resulting morality issues hit the tube, we see the abduction scenario expand to include bizarre genetics experiments, where humans are mated, against their will, with alien life-forms.

Many other equally resonant societal parallels to the abduction scenario could be concocted, but they will all lead to the same conclusion: We create and even experience supernatural events in order to cope with our own developing evolutionary abilities. The aliens themselves, like the new-type children of animé, have the ability to communicate telepathically, read one's thoughts, and use technology as if it were part of their bodies. Dr. Mack, in his demand for more research into what he believes is a valid phenomenon, unwittingly stumbled on the truth. The abduction experience goes beyond Western scientific methodology, he told the *Calgary Herald* in 1994, and "we need a more complicated ontology to grasp it." We don't need that more complicated ontology in order to understand the aliens, however. We need it in order to understand ourselves.

The rise of deeply paranoid or supernatural preoccupations in our culture, whether they're over spontaneous human combustion, chain letters, or the aliens on *The X-Files*, results from our unwillingness to embrace the technologies and abilities we have already developed. When we negate their intrinsic empowerment and retreat from the overwhelming moral conundrums they impose, we fall into hopelessness, paranoia, and delusion. We have trained ourselves over centuries to relinquish our access to technology and entrust our spiritual dilemmas to elites who supposedly know better. As presidents resign and priests abuse children, we become aware that these elites are no better prepared than we are to tackle the future. This forces us to face the inevitable. We are moving into chaos.

Yet if we begin to explore and recognize the underlying patterns in that chaos, we will be a lot more comfortable, as individuals and as a collective, tackling the issues that peek out into the mainstream consciousness as delusion and superstition. The world need look no less miraculous. For example, Oprah Winfrey did a show in 1994 dedicated to miracles. One of her guests was a woman who, "miraculously," saw a detailed image of the Virgin Mary on a flour tortilla as it baked. She attributed the miracle to God and the spirit of the Virgin, who made themselves apparent to her through her tortilla in order to invigorate her waning faith.

This woman's religious joy really came from her ability to recognize a pattern in an otherwise random system. The emergence of the Virgin's face in the cooking dough may defy the laws of probability, but it certainly conforms to the principles of recurrence and intermittency. Certain shapes recur in nature and in culture with amazing regularity. The seedlings of a maple tree are configured just like human lungs, and a household electric socket looks just like the face in Edvard Munch's *The Scream* (to me, anyway). Kids stare at clouds in order to find faces or the shapes of animals, and at the moment of recognition they shout with glee.

The reason certain images and totems become pancultural symbols is that they resonate strongly with many different people. And what is resonance? Resonance is a "sympathetic vibration." You hit one piano string and certain others, the ones we hear as harmonious, vibrate along with it, giving it timbre. The human face—two eyes and a mouth—is among the earliest images a baby comes to recognize. We naturally look for objects and arrangements to resonate with the appearance of a human face. We try to recognize the repeating patterns in nature, just as a mathematician looks for the identifiable nooks and crannies in a fractal. These are our footholds in the chaos. A miracle is simply the recognition of a pattern. We get to see that there's an order to it all.

Many of our fledgling cultural progressives, particularly those old enough to remember the Kennedy assassination or McCarthy hearings, tend to see a bit too much order to it all and verge on conspiracy theory. They have learned, usually through psychedelics exploration, occult studies, or comparing notes with one another, how to recognize some of the more subtle patterns around them, but they assume that someone or something is arranging everything to be this way. The occurrence of very similar alien sightings and abductions over the course of several decades, for example, proves that the same governmental agency has been responsible for all of them! This atheist but mechanistic worldview attempts to explain away chaos by attributing its properties to powerful yet undetectable secret organizations that have been operating in one form or another since before recorded history. Today, they are thought to remove the livers and kidneys from cattle, conduct sex rituals in the basements of royal palaces, and make up the majority of the ruling elites within the CIA, NSA, and British MI6. Their minions, the "men in black suits," abduct people who have figured out too much about them.

Angels, aliens, and secret agents all serve the same function: to explain away the spontaneous order beneath the surface of chaos. For adults, it's preferable to have an explainable nightmare than a lucid dream.

Metaparanoia

The chaos-acclimated kid laughs at the paranoid and superstitious tendencies of his elders. The screenager satirizes his parents' inablity to cope with self-similarity and recurrence by creating mock cults and collectives, distanced in shells of self-conscious media. It's camp. In doing so, he immunizes himself from the panic and anxiety of adapting to a chaotic world and shows the rest of us how media and technology can be used to demystify rather than perplex.

Among the hundreds of GenX-style mock cults publishing pamphlets and establishing Web sites, the "Church of the SubGenius," created by a group of artists in Texas, stands out as the leader in self-conscious hype and self-actualization through humor. "You may have snapped already from the information disease," warns Church literature. "Look to the High Unpredictables of the Church of the SubGenius for pan-cultural deprogramming and resynchronization! Perfect your subliminal vision . . . edit your memory . . . relive your reincarnality." On the outermost membrane of its veneer, this is a call for membership in a paranoid New Age cult. The enemies are technology, a conspiracy of thought programmers, and spiritual elitists. "Defy the sinister Star Forces which mock us all. Evil demons have kept the truth from humanity for thousands of years—God has been misquoted all this time!"

Almost anyone who comes across Church of the SubGenius literature soon realizes that this is a humorous satire of New Age devotees and self-important, nonironic conspiracy theorists. Twisting language reminiscent of a Scientology promotional flyer, the Church promises that its members will "create the highest possible earnings from the psychodynamics of abnormality," and then pushes the envelope of credibility even further by adding that it will "turn Conspiracy-implanted personality disorders around and channel them into an illusion of creativity that will fool normals and GET YOU SEX." Church dogma, as such, revolves around "Bob," who helps his devotees achieve financial and spiritual success. The object of the devotee is to do little work and slack off as much as possible. Devotion earns you slack, because Bob is the "Slack Master." The ideas of work and normalcy are being programmed into us by aliens from outer space. Church members learn through the words of Bob how to deprogram themselves and escape from the conspiracy.

On a deeper level, participation in the Church of the SubGenius truly does function as social deprogramming. The entire enterprise is a mission of slack for its creators, who bring their weirdest dreams and abnormal thoughts to the surface for fun and profit. They have absolutely exploited the "Conspiracy-implanted" disorders and "channeled them into an illusion of creativity," just as their literature instructs. The Church of the SubGenius is a group of creative but weird slackers who sell their abnormality. Ironically, the Church delivers exactly what it promises: deprogramming and slack. By enjoying the satirical literature, we distance ourselves from the all-too-serious and needlessly mystified spiritual pretensions of our "real" churches and employers. Spiritual fulfillment and gainful employment are crocks—at least in their current form. The Church of the SubGenius offers an alternative to abuse by those whom we allow to maintain authority over us.

The best efforts of media-based mock cults work like performance art on their consuming public. Each pamphlet, Web site, or 'zine works only within the overall context of our current, mystified perception of spirituality and technology. They imitate the thought constructions of conspiracy theorists and frightened New Agers, magnifying them beyond cliché and into the stark clarity of absurdity. They combine so many of the features of our current climate of spiritual paranoia, and do it using so many levels of transparent manipulation, that they generate a turbulence of concept and image. Distanced by comedy, we can appreciate this collage of absurd notions for its underlying order—an order held together by our own superstitions and blind allegiances.

A number of new, slacker-inspired media cults work even more directly to stoke our ignorant fears into self-aware comedy. Nesting itself precariously between paranoia and satire on one side, art and graffiti on the other, the "Schwa" project both inflates and deconstructs our obsession with alien abduction. In a series of stark line drawings using stick figures, graphic artist Bill Barker parodies almost every possible permutation of the abduction or invasion fantasy, from flying saucers attacking urban zones with laser beams to giant alien factories processing humans down to their component parts. He even reinterprets classic conspiracy scenes, like the Kennedy assassination, as alien-inspired scenarios (the second gunman was in a flying saucer!). Aliens are represented in simple iconic versions of the large-eyed abductor-examiner so common to the descriptions of UFO encounters.

By reducing these scenes of fear and despair to their most simple, black-and-white, graphical components, the artist strips away the distracting, high-tech, sci-fi quality of our obsession with space invaders, and shows us what it is we really are so afraid of: the emotion-numbing sterility and thought-deadening isolation of the modern experience. The *schwa* itself, that upside-down letter ə signifying the indeterminate vowel sound in the first syllable of a word like *alone*, also represents any strange or unexplained occurrence. The unfamiliarity of the world around us is reduced to a single icon.

In addition to comic books filled with graphics, Schwa puts out a line of "alien defense products," including alien-detection key rings and "instant stick person" necklaces that are to be used in the event of an alien-related crisis. They are purported .to be coated with space-age substances like "xenon," which, like the tags worn by workers in nuclear power plants, supposedly change color to indicate danger. The objects are worn by Schwa fans as mock-talismans, satirically warding off unknown forces while commenting on how dangerous the modern experience has become for us both in reality and in perception. But, on a subtler level, the little tags and necklaces *function* as talismans, too. They emblemize an awareness of our real discomfort over the increasing unpredictability of the world around us. The symbol of a tiny alien or schwa doesn't really have the magical function advertised on its packaging, but it does condense the diffuse anxiety of our age into a single, self-consciously worn icon. It transforms the mysteriously nebulous into something palpable, wearable, and even laughable.

Maybe that's how talismans worked all along. It's not size of the wand nor the magic in it, but the anxiety we project onto it.

Techno-shamanism

Screenagers are not without their own more sincere efforts at reducing spiritual anxiety. They are fully aware of the underlying causes of our malaise: a deep-rooted fear of technology and the intimacy that it portends. By accepting the notion that technology can play a part in the forward evolution of humanity toward its greater spiritual goals, the children of chaos exchange the adult, paranoid response to the impending colonial organism with a philosophy decidely more positive: *pro*noia.

The rituals and play of children of chaos aestheticize the process by which human organisms use technology to come together into a single networked whole. For example, when the Power Rangers summon the ancient spirits of dinosaurs, it is in order to promote themselves to the next evolutionary level. The dinosaurs appear as robots called zords, which can then combine, with the Power Rangers inside them, into a giant gundam robot. Just as each stage of evolution leading to human organisms involved the coming together of component parts into a single networked whole, the kids watching *Power Rangers* are learning that for humanity to face the obstacles

154

ahead, they will need to develop, somehow, into a greater coordinated being. This is our greatest social and spiritual challenge, and the children of chaos address it with the optimistic faith of zealots.

Ritualized imitations of colonial organisms are emerging everywhere in the chaotic culture of young people. The mosh pit is one of the more obvious attempts at creating an organismic sensibility in a social setting. Rather than dancing with partners or in formally designated lines, the kids in a mosh pit flail about with abandon, crashing into one another and climbing over each other's bodies. It is a Dionysian sea of limbs and sweat. Some leap off the stage onto waiting hands, to be carried over the crowd until they fall into the mass. Others slam into each other—sometimes very hard. It's not uncommon to get injured, but that's not really the object of the dance.

The mosh is extremely sexual. It began as an all-boy sport called slam dancing during the late-punk period of the 1970s, and although it was understood by the macho young skinheads as an aggressive male rite (are you man enough to mosh at CBGB's?) the animalistic intensity of a hundred leather-clad adolescents crashing against one another all night, if not exactly erotic, was at the very least sensual. But the kids were not overtly reaching for one another's private parts, and the eros of the dance took on a tone very different from a group grope. It was entirely subordinate to a greater aspiration of communal bonding. For however macho and violent the slam-dancing appeared to outsiders, it was actually an intricately orchestrated cooperative effort.

Today's mosh is less personally aggressive than an old ladies' tea party. There is no backstabbing or social climbing. This is pure participation in a group revelry, and its surface machismo is part of what allows it all to happen. The boys are slamming each other so hard and with such apparent aggression that no one would accuse them of "getting off" on touch-

155

ing one another. There is no sentimentality or softness that could make an average teenager question his masculinity. But the level of cooperation necessary to achieve a successful mosh pit is astonishing. Fallen moshers are immediately helped to their feet by their neighbors. Most of the participants wear heavy boots, but despite the vigor with which they stomp about, no one is to step on anyone else. Moshers have developed elaborate maneuvers allowing them to fling themselves up on the shoulders of other dancers, who pass them around over their heads. The mosh has become such an accepted form of expression that it occurs at nearly every rock concert, rather than just at hard-core downtown clubs. It is now commonplace for girls to join in the pit, too, and at big concerts like Lollapalooza organizers distribute flyers detailing mosh etiquette and safety suggestions.

The popularity of moshing is attributable to the overwhelming urge of groups of people to move into an organismic relationship to one another. Just like the "wave"—a stadium happening that rose in popularity during the 1984 Olympics, where fans all stand and wave their hands sequentially around the arena—the mosh pit allows participants to experience one another as part of a larger organism. Get any large group of people together at the end of the twentieth century, and they will find a way to experience their group dynamic, whether as a cooperative performance event or, as in the case of more patriotic, dualistic sports events, a mass riot. Although the expression can be violent, the urge is the same: form a networked whole. At least the kids are doing it without intentionally hurting one another.

As in a dynamical system, the turbulence of the mosh allows for unique relationships to form. Although they are not building bridges or conceiving physics theorems, the dancers are deeply involved in a group enterprise: the creation and sustenance of a resonating organism. For a moment, and maybe even several hours' duration, the dancers no longer experience themselves as isolated individuals performing cause-and-effect actions with single partners. They are in a group relationship, with extreme sensitivity to one another's needs and moods. Each dancer is a potential high leverage point who can change the rhythm and flow of the entire mosh.

The sexuality of the mosh pit goes beyond simple one-on-one lovemaking. It amounts to a complete dissolution of personal boundaries, as the bodies of countless strangers press against one's own. A fist in your side, not bone-crushing but hard, followed by a chest in your face and elbow to your head, all delivered with passion and random rage let you know that you are part of a teeming mass. The force of the interaction makes up for the lack of standard intimacy. Each dancer is a cell in the total mosh organism. Without sacrificing any individuality, each dancer becomes part of a greater being. The violence itself amounts to a kind of test: Even allowed to express our rage and angst, we won't hurt one another. The total release of so many individual wills yields an intensely agitated, close-to-the-edge surge of high-impact contact, but it almost never turns into a brawl.

The mosh pit is only one of many attempts by the children of chaos to approximate the dynamics of a colonial organism, which may very well be the next necessary phase of human evolution. A colonial organism, like a coral reef, is a community of individual creatures who form a networked whole in order to promote their collective survival. A coral reef is made up of millions of tiny organisms who function individually as separate creatures, but collectively as a community. They benefit both from their ability to behave as individuals in certain circumstances, and as a unified whole in others.

Most people loathe the colonial organism as a model for human society because they fear the accompanying loss of individual freedom. No one, especially an American, wants to be part of the herd. But a colonial organism does not require that its members sacrifice their personal will. In fact, when a group of individuals is networked, each member has more, not less influence over the entire group. Thanks to a property that chaos mathematicians and biologists called "phase-locking," human beings can experience the benefits of colonial living without an accompanying loss in individual will or personal influence.

For example, if a group of women move into a house together, their menstrual cycles will eventually coordinate. This instinct has not evolved to curtail individual expression among cohabitating females. On the contrary, when each is in the same part of the cycle at the same time, the entire group is more, not less, sensitive to the state of the mind and body of each member. This is a phase-locked system. Just as a group of people at a factory who work and sleep on the same shifts will develop an almost mysterious rapport, in a phase-locked system information will seem to pass from member to member without any overt communication.

The more phase-locked a system is, the stronger its ability to maintain itself. The human body has thousands of processes that are phase-locked to one another, as does an ant farm or an entire ecosystem. Phase-locking is generated by cycles like the rotation of the earth, day and night, seasons, heartbeats, breath, or birth and death. Just as the repeated cycles of nonlinear equations generate fractals, the repeated cycles in the real world generate fractal communities—colonial organisms, which exhibit self-similarity, turbulence, remote high leverage points, and chaos. Phase-locking is not goose-stepping. It cannot be designed or enforced, but must arise naturally out of the needs and tendencies of the entire group. The likelihood of phase-locking arising in a group is increased when the group is networked together. A group of kids in a mosh pit will experience something much closer to phase-locking than the same group doing a line dance. In a disciplined line dance, the number of interactions is limited by design. The steps are memorized, and the movements proceed in a linear fashion. A mosh pit is a free-for-all. The dancers use randomness and violence to break through linear social convention in order to reach a state of turbulence. The patterns within that turbulence form the basis of the phase-locking that is the real goal of the activity. It creates the feeling of a colonial organism, and to these kids that feeling is a good one.

A more advanced version of the mosh pit is the rave, the dance event consciously and intentionally designed by kids to promote the colonial experience in a purposefully spiritual context. Rave kids attempt to phase-lock as a community by sleeping during the day and dancing at night, joining together in an alternative circadian rhythm from the overculture. The rave itself, unlike a regular nightclub event, usually occurs outdoors in a remote field, or in a giant abandoned warehouse. Several thousand kids, many of whom have taken mild psychedelics, make their way to the event and then dance all night long. The electronic music throbs at exactly 120 beats per minute, the fetal heart rate. Lights and lasers flash. The music itself is structured like a comic book and made up of extremely discontinuous samples of other musics and cultures. African and Native American tribal drums might be juxtaposed with sounds from a video game or traffic jam. This is a conscious manipulation of sound in order to create the effect of cultural and historical compression: the sonic rhythms from as many cultures as possible are combined so that kids can dance and trance together on many different levels at once. The entire event becomes fractal in structure and even surface expression, as images of fractal designs themselves are often projected onto walls or screens and worn on T-shirts. Videos of computer-generated chaotic forms play on screens and monitors, while in the quieter corners, young intellectuals share their thoughts on the emerging chaotic culture.

The rave is self-consciously phase-locked, and self-consciously technological. Ravers use their computers and digital recorders to create their music and design their graphics. They embrace technology for its ability to sample and recombine sounds and images from throughout cultural history, and even more for technology's ability to forge a new global culture. Rave philosophy, as outlined on the back of a clothing label or "hang tag" (one of their main conduits for written communication) generally credits technology with promoting global community and cultural tolerance. Their justifications are fairly straightforward. Aspects of different cultures can be sampled, overlaid, and experienced in synergy with one another. Computer networks and global media promote communication among people who may not otherwise have come into contact. Like-minded individuals achieve new lines of connection and can converse in self-selecting communities on computer bulletin boards, Internet newsgroups, or e-mail.

Rave dogma, as such, is based in the Gaia hypothesis of James Lovelock, which looks at the earth as a single organism made up of a number of phase-locked communities. Human beings serve as the neurons in the global brain, and technology allows separate neurons to link up to one another and behave in a coordinated if chaotic fashion. Once we are fully networked, according to the rave kids, we will achieve a global consciousness. Humanity reconnects to its mother planet and learns to promote her survival rather than selfishly exploit her resources. Further, this does not amount to a retreat into tribal, unconscious herd behavior, but a leap forward into metaconsciousness. We learn to think, breathe, and act as a tremendously networked planetwide organism.

Wide-eyed adolescent optimism? Of course it is. The ravers will be the first to fully endorse pronoia as a working model of reality. Still, the experience of the rave offers a glimpse of true group consciousness. Disjointed samples from multiple cultures yield discontinuity. The discontinuity generates turbulence. Meanwhile, the many rhythmic elements and synchronous life-cycle decisions create the opportunity for tremendous phase-locking and group identification. It climaxes in a frenzied, thousand-person, psychedelics-enhanced dance event, in which an extended moment of group bliss serves as an intensely powerful method of self-imposed social programming. The ravers want to convince themselves that this next phase of human evolution is, indeed, a possibility.

Whether we as adults choose to partake of this revelry or not, it behooves us to at least entertain the possibility that technology and the turbulence it promotes stands a chance of restoring rather than further distancing us from our spiritual natures. It would mean dispensing with a preordained but hierarchical model of the world and learning how to accept an existence that is ever-changing and fundamentally fractal.

4

THE FALL OF GRAVITY AND THE RISE OF CONSENSUAL HALLUCINATION

"Hey! You guys are upside down!"

Commander of the U.S. space shuttle *Atlantis* to
Russians aboard the space station *Mir* as hatch opened
in that historic docking of 1995

We love stuff because we can touch it. It's real. It has mass. It falls to the ground when we let go of it.

There's also a special value that real objects can take on, too. A violin played by Stravinsky, a pen used by James Joyce, or a pillow on which the Buddha slept are treasured more than identical objects from the same periods because of the sentimental value we project on them. Their molecules are no different even if their price tags or museum placement are. The "genuine article" is an ethereal notion, at best.

Traditionally, for objects to take on magical significance

they must be connected to history and lineage. The transubstantiation of a wafer into the body of Christ doesn't happen in a person's living room, even if he says all the right prayers. It only happens when an ordained priest conducts mass. A samurai sword, Fabergé egg, or original copy of the U.S. Constitution is passed on from one person or institution to another only when the recipient has demonstrated proper allegiance to and regard for the significance of the object, either through application, ritual, or a huge outlay of cash at Christie's. No matter how much Westerners might complain about Japanese investors scooping up Picassos, the buyers are only conforming to the system by which we prostrate ourselves in modern culture.

The forms of paganism kids explore today are all ways of reclaiming the right to assign value to the physical. A blessing conducted in a back alley can turn a trash can lid into a holy platter. A kid making skateboard stickers with a Magic Marker on Avery labels can claim the same level of authenticity as Chagall. If he's accepted as readily as Keith Haring, his work may even become as highly valued.

This desire to generate or declare objects of value is only an intermediate stage in the progression toward chaos. It is a reclamation of a certain power, but the surrender of another. The quest for an authenticated, factory-sealed, original edition, name-brand Pog reduces its buyer to the level of a parishioner among priests. He can only obtain the object of his desire by standing at the bottom of a supply chain. Kids who have opted out and begun generating their own totems and talismans imbue them with "magic" in order to fight back. They see the established system of evaluation as true tyranny, and act in order to counter one force with another.

The way to break the spell is to drop the idea of authenticity altogether. Unless an object is irreproducible, like an original sculpture or painting, authenticity is just another form of static elitism. It's a purely symbolic gesture using an object's pedigree to maintain the lineage granting it. As we move into an electronically mediated world where symbols need have no mass, the rules of the game change. What is an "authentic" Nirvana CD? The first pressing? A signed copy? Does that make the music any different? No.

The physical world is authentic by its very nature. Maybe we didn't realize this until we all started thinking about "virtual reality," but every time we breathe air, look at a tree, or make love, we are interacting in authentic flesh and blood. It's real. When we interact with icons and representations, we have moved into a different realm. It need not be confused with authenticity at all. A particular crucifix may have mass, but the symbol of a crucifix does not. No, I can't own a piece of the True Cross, but I can make my own. With or without authenticating ritual, the symbol and its resonance are the same. A rose by any other name would still smell as sweet— but without consensus language we couldn't share the thought with anyone else.

The deliberate confusion of symbols with objects keeps symbology itself—perhaps the most empowering and magical of human abilities—out of our reach. As long as symbols are represented by real objects, their power can be held. Because physical objects are subject to the force of gravity, they can only fall down. Superiors in the objectified hierarchy can "dub" their inferiors. They hold the scepter and the crown. That's why they sit up on thrones and altars: symbolic power, when it's attached to objects, only comes down from above.

When symbols of power are divorced from objects, they are also divorced from gravity and everything that goes with it. In contrast to an object, an image can be copied and distributed over electronic networks instantaneously. Entirely new rules of commerce have been developed to cope with this, ranging from copyright and patents to royalties and residuals. But lawyers and litigation are just a last line of defense against the impending turbulence. Ultimately, the move into massless commerce, religion, politics, and ideological warfare makes 90-pound weaklings indistinguishable from the heavyweights.

Media Space

Media used to be a top-down affair. A few rich guys in suits sat in offices at the tops of tall buildings and decided which stories would be in the headlines, and how they would be told. However much the printing press and literacy movement empowered the masses to absorb information, the high price of manufacture and distribution kept the production of that information in the hands of those who controlled the equipment. The likes of William Randoph Hearst dominated the content of our media and directed public opinion from the top.

As print production became cheaper and more accessible thanks to typewriters and photocopying, news and storytelling moved to television and radio. Again, because these new, electronic media were so expensive to operate, not to mention obtain broadcast licenses for, content could easily be controlled by a few rich men and their even wealthier sponsors.

As a result, we came to think of information as something that got fed to us from above. We counted on the editors of the *New York Times* to deliver "all the news that's fit to print," and Walter Cronkite to tell us "that's the way it was." We had no reason not to trust the editorial decisions of the media managers upon whom we depended to present, accurately, what was going on in the world around us. Television proved a better programmer than print, too, because it demanded even less of its audience. No reading, no distracting page turning, no freedom to choose articles or the order of presentation. Just tune in, sit back, and zone out.

But by slowly shifting our sense of trust onto the electronic image, those who hoped to maintain control over the content of media created the conditions by which they lost it. We loved TV, and as more and more of it was provided to us, those terrific properties of turbulence led the tube to promote chaos over anything else. For electronic media doesn't intrinsically empower an elite of content providers the way printed matter did. It doesn't obey the laws of gravity.

Cable television and home video technology combined to make the world of television a less ordered place. Much to the chagrin of network news advocates, CNN and CSPAN rose to prominence and broadcast live news footage from around the world with much less editorial oversight. Television began to serve the purpose its name indicates: it allowed us to see what was going on somewhere else. Its role changed from a source of passively absorbed programming into a tool for remote viewing. Well-meaning liberals, like *Los Angeles Times* columnist Tom Rosenstiel, were horrified by this transition, writing in a *New Republic* cover story that CNN "has even had a pernicious effect on the rest of journalism: it has accelerated the loss of control news organizations have over content." I'd agree except for the word *pernicious*. As centralized control over news agencies erodes, the content of media becomes much more responsive to the needs of the world culture it serves. Rosenstiel argues that CNN "gives voice to political leaders who otherwise lacked political standing."* What's wrong with that? When a dictator with a million sword-wielding would-be martyrs wants to tell the world something, I'd rather he do it over CNN than a mound of dead bodies. Supply-side news control breaks down when international networks are willing to point their cameras anywhere.

Public demand gets more of a voice in decentralized mediaspace, too. Liberals ironically fearing the will of those they would protect—the undereducated "masses"—claim that sen-

*Tom Rosenstiel, "The Myth of CNN," *The New Republic*, 22 and 29 August 1994.

sationalism will replace wise journalism as the news stoops to meet consumer appetite. Tough. Stories are sensational for a reason. Rather than numbing ourselves to what titillates, we should take a good hard look instead. In the United Kingdom, sensationalist tabloid media took responsibility for eroding the stature of the monarchy in the minds of its subjects. Only the tabloids were willing to appeal to the public hunger for royal sex scandals, Camillagate, and Princess Di's bulimia. These bottom feeders exploited the broadly experienced but largely unstated perception that the royal family was no better than anyone else. Had they been elected rulers, their personal transgressions could be forgiven a bit more easily because their positions would have been earned based on certain qualifications. People resent being the subjects of those who were born "superior," especially when they are no more worthy of their status than anyone else. There was no moral high ground.

Traditional media outlets, out of deference to the throne and fear of treason charges, never reported on the moral transgressions of the royal family. Prince Charles and Lady Diana were not the first royal couple to engage in extramarital affairs—our modern-day tabloid media simply had the guts to report it. Once the "lower" forms of media began to include formerly treasonous repudiations of the royal family, the more traditional news media ran stories and editorials considering the issue of whether the monarchy was serving any useful purpose in modern-day Britain.

This trickle-up quality of media reverses traditional supply-side information economics and threatens those who would keep obsolete social institutions in place, especially when those institutions are the media content manufacturers themselves. In order to keep up with their tabloid competitors, network news shows "stooped to their level" and raced to appeal to public appetite rather than particular sponsorial agendas. The overriding agenda of the advertiser, after all, is to have viewers. The ratings of lower-quality tabloid television have begun to decline on their own because the more fully staffed networks have adjusted their programming to suit an audience that demands the truth.

This is better for us. In the 1930s, the press spared Franklin Delano Roosevelt and the American people the humiliation of learning that the president had been permanently crippled by polio. He was never photographed in his wheelchair, but was instead pictured swimming or engaging in other physical activity that did not require the use of his legs. Whom did this serve? What did this do for the millions of physically challenged people in the United States who continued to think of their conditions as shameful? If it takes a tabloid media acting on titillation and shame to expose the dangerous high voltages and pernicious denial in our culture, then so be it.

When cable and tabloid fall short, public access television and call-in radio pick up the slack. However horrific a G. Gordon Liddy's radio show may get, it is merely a long-overdue response of a public against what it experienced as extreme domination. If our culture in its former stage of top-down mediation could be likened to the passively absorbing fetus or breast-feeding infant, then this second stage can be seen as the "terrible two's." When a baby first learns to speak, he soon realizes that his words can have an effect on those around him. Lacking a plan of his own and the words to express it if he did, the child simply repeats the one word that can so easily negate everyone else's: *no*. The "hate jockeys" and their equally vehement callers are people who still don't realize that there's no gravity in the datasphere. They see authorities at the top of every social network, and scream at these parent figures for relief, independence, and out of plain old frustration. Not to diminish the human suffering of a Waco or Oklahoma City, but the underlying impulse is more akin to an angry child smashing his mother's favorite piece of crystal than a well-thought-out or supportable social outcry. Terrorists and extremists like those at Oklahoma or Waco want *no* government, because they can no longer conceive of any organizational structure, even one based on organic reality, that wouldn't be turned against them. As one "patriot movement" member explained after the Oklahoma bombing, "Nothing in government occurs by accident. If it occurs, know that it was planned that way. To be planned, there must be planners. If you have planners, you must have a conspiracy."* This faulty syllogism is the same as that of a child who can't understand his parents' seemingly random actions except as some form of conspiracy against himself. "I won't do it again," one nationally famous six-year-old victim of a child custody battle told his mother as the court turned him over to a biological father he had never met for reasons he could not understand. It's easier to believe we are being consciously punished.

The paranoid hate-mongerers are more sadly damaged victims of the hierarchical worldview than those who passively absorb everything their media tell them. However, like a foot that wakes up from a numbing frostbite, there's still a lot of pain and sometimes even the amputation of a few toes before it can be brought back to life. As we restore what has been reduced to a unidirectional nervous system back to its full, multidirectional communicative ability, we will be forced to learn just what condition we are in.

The real question, then, is how do we bring ourselves to the next stage of development as quickly and painlessly as possible? It is to be accomplished by understanding what a world without hierarchy looks like, and learning how to navigate it. The number of channels, voices, camcorders, and, finally, home computers is fast bringing our mediaspace into a state of full turbulence. It is becoming fractal in nature. Our

*"Demons and Conspiracies Haunt a Patriot World," *New York Times*, 31 May 1995.

job is to look for the underlying patterns and natural organic forms breathing within it. As media expands to encompass all of us as both content consumers and content providers, its intrinsic weightlessness becomes undeniable and even potentially liberating. As always, it's the kids who are best preparing themselves for this impending loss of gravity.

Weightless Training

Pinball is a battle against gravity. After we whack the silver ball with the springed plunger, it's only a matter of time before it rolls back down the incline plane toward us. We stave off the inevitable with precise flipper flutters, but eventually the force of nature wins out and the ball rolls back down the shoot. Extra games, however delightful, are just booby prizes. Like the promise of reincarnation offered by the Hindus that can't really be enjoyed because it's *someone else* who comes back, each new ball is like starting over and only exacerbates the anxiety that this ride, too, is only temporary. Sooner or later, it's going down the chute.

The evolution of the video arcade games that replaced pinball was a move away from gravity-based play and toward weightlessness. Think back to the first time you tried Pong, the original video game from the mid-1970s. The two white lines representing paddles and tiny white square representing the ball didn't move or feel like objects from the real world. There was no impact of the ball against the paddle. The ball hovered weightless as it traveled from one side of the screen to the other. The bottom of the screen wasn't really "down" nor was the top "up." No matter how much you moved your body in sympathy with the ball, only your fingers on the control knob had any effect.

This difference in experience holds the key to understanding how to orient oneself in a weightless world. Equally important, the direction of the developing video game culture and aesthetic indicates what we can expect as our mass media and culture at large follow their lead. The advancement of video games over the past three decades was based on the emergence of new technologies. It was less a consciously directed artistic development than a race to utilize new computer chips, imaging techniques, and graphics cards. Every time a new technology arrived, game developers would redefine the essence of their game around that new hardware.

At each successive leap in video game development, then, there is a return to technology and its underlying nature. Even if two or three years pass without a technological innovation, the games can only develop so far on a particular platform before a new platform is introduced that redefines the medium. Each new console, whether it is Super Nintendo or Sega Saturn, comes equipped with a "killer app"—a piece of software (the game cartridge or disk) that makes use of the new hardware and can't be played on more primitive machines. Super Mario was designed to take advantage of the scrolling capabilities of Super Nintendo (1993); the little character runs across moving, multilevel backgrounds that scroll by with unprecedented clarity. Virtua Fighter (1994), which utilizes the "polygon engine" inside the Sega Saturn, has fighting characters made from rendered three-dimensional polygons. The style and content of the games is based on the specific qualities of the new machines as they become available. In this way, technology itself enforces the direction of video game evolution.

In one sense, the games don't change at all. Since the very first video games, there were only three main archetypes. These have been embellished over the years, but, like modern plays that follow the principles developed by the ancient Greeks, they still hold the essential mythic structure of almost every further-developed scenario.

There are the "duel" games. Pong was the first. Two players

hit a ball back and forth trying to get a ball by their opponent. The skill is defending your side of the screen and penetrating your adversary's. The Odyssey game console's simple processor and basic software allowed only for monochromatic representative shapes on the screen–short lines for paddles and a dot for the ball. Not many other duel games could be articulated with such simple shapes. The only possible point of view for the players was directly above the playing field, in an essentially two-dimensional representation that did not move. No aesthetic considerations were made. The paddles did not look like anything except place-holders. The player was required to imagine the real-world qualities of the scenario.

The next major evolutionary leap of the duel game took place once video game machines gained true graphical capabilities. Now that the player could be represented by a drawing rather than a simple line, the point of view changed from above to the side. Characters became cartoonlike two-dimensional images of battling opponents. Scrolling technology permitted the "camera" to move, as if on tracks, to the left or the right, following the dueling characters back and forth. Street Fighter (1992) utilized these techniques the best, and players fought each other directly, Rock'em Sock'em Robot style, in full color on the television screen. Mortal Kombat (1993), based on the same technologies, used digitized photographs instead of cartoon animations, which gave players more violent-looking finishes to each round of combat.

The latest stage in duel games, so far, brings three dimensional polygon engines into use (1994–1995). Now, the characters themselves are rendered in 3-D, so that the "camera angle" can change not just from side to side, but all the way around the characters, below them, and above them. Players no longer have "sides," as they did in Pong or in the Street Fighter games. The ring itself is circular, and players move about freely, defending only the three-dimensional limits of their characters' bodies.

Another evolutionary path, the quest games, began in pure text, on the computer. The player wanders through a simulated world, exploring rooms, picking up objects, making choices, and trying to figure out what the hell is going on. Eventually, the player's task becomes clear, and he uses what he has learned about the world in order to complete it. The quest could be as simple as rescuing an imprisoned princess, or as complex as clearing oneself of murder charges before the cops or mob find you. These games are information heavy, and make best use of a computer's huge memory, rather than a video game console's fast rendering and joystick controls. The game Adventure was one of the first of this genre (developed through the early 1970s) and worked purely through verbal descriptions and typed commands. Paragraphs of text would tell the player that a goblin had appeared or castle was on the horizon, and the player would then type what he wanted to do: "punch the goblin; walk to the castle gate." Slightly more advanced text-based computer games, like Net Hack, used regular ASCII-type symbols to create graphics (ASCII graphics). A person might be represented by an @, and a wall by a series of dashes. Players would use a combination of cursor (keyboard arrow) movements and typed commands to move through the adventure. As in Pong, the world was observed from directly above.

The arrival of Odyssey and the other dedicated game systems (late 1970s) brought slightly more graphical but far less strategic and detailed versions of the quest. In a game called Haunted House, simple lines were used to represent walls and doorways, keys and cups, people and dragons. Although the player could move through the world more easily using his control pad than he could by typing, there were many less decisions to make and less possible outcomes. Once Super Nintendo and Sega Genesis scrolling became available, revamped games like Zelda emerged, which offered much more detailed cartoon graphics of the world to be explored.

Instead of the player viewing himself from directly above, he sees his character from high above and slightly behind. Like a city map where the buildings are all drawn in perspective coming up from the plane of the page, this "above-and-behind" perspective allowed us to see a bit more of what the game world looked like without losing our sense of the overall map. Still, because of its limited data storage, the video game console could not provide as complete a world as the computer.

Back on the computer, quest games continued to take advantage of the PC's immense data storage capacity, and slowly improved the games' graphical capabilities. The games allowed users to move from scene to scene, as they explored the world, and a single screen shot was used for each location. The user clicked on icons or typed text in order to take actions or move. Each game had to come up with a reason why players should explore its world. In Deja Vu, a popular game from this genre and period (mid-eighties), the player has woken up in a bathroom stall with total amnesia. As the user learns who his character is, he learns about the world of the game and his predicament.

Later forms of this genre took their cue from Myst (1993), a CD-ROM game for the PC that relied even less on icons and text, and moved the player into the pure point of view of the character he's playing. Back on the ground and through the eyes of the character, the player clicks on what he wants to do and where he wants to go. Although Myst hardly launched the paradigm shift the industry magazines predicted, the game popularized the quest scenario by making commands invisible and creating a fully rendered, first-person interface.

The evolutionary path of the quest and duel games was the same: they moved from overhead-view, iconized graphics to side-viewed cartoon graphics and ended with rendered three-dimensional worlds seen from close to the point of view of the character.

"Apocalypse" games progressed through the same developmental stages. In this archetype, the player, alone, battles successively more challenging waves of attackers. In the first games, like Asteroids and Space Invaders (1977), the player remained fairly stationary, or moved along the bottom of the screen, as an onslaught of projectiles assaulted him. The action was all observed from above, and the simplest of line drawings were used to represent alien ships and approaching asteroids. Pacman and its even more popular successor Ms. Pacman (early 1980s) used full-color monitors and gave the player the ability to move. The action was still observed from above, but the player could maneuver toward or away from the attacking ghosts and ghoulies.

Mario, Super Mario, and Sonic Hedgehog took advantage of the scrolling technology of early-1990s SuperNES and Sega Genesis (as well as the late-1980s arcade implementation of this technology) and moved the player down to the side view. Now, instead of just moving through one map at a time, the background scrolled by as the character ran toward the right side of the screen. Hundreds of other games followed this formula, both in arcade and home versions, from Mega Man to Haunted Castle.

The evolution toward 3-D perspective games moved through the "above and behind" stage and eventually emerged as character point-of-view (POV) adventure Doom (1993). The frighteningly revolutionary aspect of the Doom games, no matter what the pundits say, is not that they are more violent than any games before them. They are simply more real. The player is completely within the point of view of the character, who must battle his way out of one hellish nightmare scenario or another. In one game, the player is in hell, in another he battles his way out of a Nazi prison, and in another he is on a space station that has been overrun by evil aliens. The world is in complete disarray, and the player must defeat and overturn the status quo or die trying.

The effort by concerned parents and politicians to make video games less apparently violent is earnest but misguided, because they do not understand the underlying reasons for the increasing realism of these games. Social scientists have sought to demonstrate a causative link between television or video game violence and aggressive behavior in children. Unfortunately, social scientists do not follow the same stringent methodology of real scientists. For example, one study shows that kids who are called into the principal's office for aggressive behavior are more likely to say that *The Mighty Morphin Power Rangers* is one of their favorite shows than kids who haven't been caught in violent behavior. That is not the demonstration of a causative link. It's akin to saying that because a higher proportion of African-Americans frequent Kentucky Fried Chicken than white Americans, eating fried chicken will make a person turn black.

A world in which toy guns must be colored fluorescent green is not so dangerous because of the preponderance of toy guns. Toy weapons are only dangerous in a world so violent that we can no longer tell the difference between the fake and the real ones. If anything, the repression of fictional violence and the confusion of adults over the differences between ritualized play and bloody murder are what lead to the actual violence committed by kids and the adults those kids become.

The true purpose of play and the violence associated with it is revealed when we examine the natural direction that video games have been allowed to evolve toward over the past few decades. There are startlingly few examples left of unfettered evolutionary activity in modern culture, but video and computer games, because they developed along the same natural evolutionary path as technology in a reverse-engineering similar to that of animé and gundam cartoons, show us the more organic impulses beneath the surface of mediated play.

In each case of an archetype's development, the games progress from objectified viewpoints to increasingly participatory ones. They turn from stories told or observed into stories experienced. Video games, like most fictional media, are an imitation of dream space. The world is generated, on the fly, by the game console as we move through it. In some games, you can even see the scenery being rendered as you move toward it. But, like dreams, the scenes are from a weightless reality. A real ball never descends an incline plane, nor does a real Nazi ever fall to the ground dead. As our tolerance for the reality of dream-death increases, we can accept more and more realistic and riveting portrayals of violent events. This doesn't make them any closer to flesh wounds—only to a more consciously experienced catharsis.

As Jung would tell us, the archetypal struggles in dreams remain the same, even if the symbolism changes from era to era and culture to culture. In video games, the central conflicts and universes remain the same over time: Our world is being attacked; I am in a struggle against another individual; or I must accomplish my quest. These are the same structures underlying dreams. But if a person goes to a psychiatrist because he is having problems in life, does the doctor try to change the patient's dreams? No. He gets the patient to remember more about them, or even dream consciously in the form of guided visualization. Dream deprivation studies have shown that if a person is not allowed to dream, he will develop psychotic delusions—hallucinations in waking consciousness. The same is true for cultures. If we deny ourselves or repress our cultural dreams as they express themselves in our media, we will experience cultural hallucinations like paranoid conspiracy theories, New Age magical thinking, UFO abductions, and more. We should not try to change our world by changing or eliminating our dreams, but we can look to our dreams for answers about why we do what we do in real life.

The unique opportunity offered by a mediated dream space is that we all experience the same dreams together. A particular game becomes popular because it offers a dream in which many kids wish to participate. Should we fill a child with shame because he has a violent dream? No more than we punish the Shaolin priest for practicing martial arts or reading *The Tibetan Book of the Dead*. Nor should we condemn them or ourselves for participating in violent game play in the weightless realm. Unlike boxing, no one really gets hurt. In mediated play, like no other, we can push ourselves into ultra-violent, physically impossible acts of aggression, and everyone can live to tell the tale. Most video game consoles come equipped with modem ports, so that players can find opponents or co-combatants anywhere in the world. Kids will wander through the corridors of Doom together, teaming up against the monsters. Most computer quest games also have networking capabilities, so that four or more players can work together or against one another over the Internet. It is as if video games comprise a technologically realized collective unconscious. A shared dream.

Unlike most computer games, though, in the act of dreaming, the dreamer gets to create the world he inhabits. The level of violence and passivity is wholly determined by the dreamer's own mind. Maybe this is why the latest and fastest-growing segment of the computer game market is simulation games, or what we can call the "God" archetype. In these games, the player develops and controls some sort of world and makes decisions about what kind of world it is going to be.

One of the first popular games of this type was called Balance of Power (late 1980s), where the player acts as the chief of state of either the United States or Soviet Union and attempts to avert nuclear war without compromising the interests of his nation. The game is programmed so that the more violent or aggressive the player gets, the more violent the rest of the world becomes. The game looks like a map of the world, and players click on countries or use menus to take actions.

The next and most popular variety of this archetype so far is Sim City (first developed in 1987). Rather than taking charge of a world already in progress, the player starts from scratch, bulldozing terrain and placing roads, power plants, homes, and industry, then watching and making adjustments as cars fill the streets and a population inhabits his town. If he does a good job, his popularity and tax base remain high. He can grow his town into a city and even build an airport or stadium. The object of the game is not to "win," but to develop a sustainable society. Although the original version of Sim City used icons and is viewed from above, newer versions allow the player to experience his city on almost any level. He can go to the street and hear the comments of his citizens or up to the sky to see the traffic patterns. This fractal approach to the God game seems designed to demonstrate the fact that the tiniest of interactions reflect the largest concerns.

For networkers, a giant God game called Civilization (1991) allows players to pick and develop a civilization as it evolves from tiny tribes to modern nation states and beyond. The players are in competition with one another for continents and power, but whoever manages to survive through the modern era, gets to cooperate in the construction of a spacecraft that moves humanity off the planet. Civilization is shared world building, and thus shared dream construction. Each player has a say, but the world of the game is determined by the consensus of the whole group as expressed in their actions.

So the dream our kids want to dream is a collective one. Through technology, they gain the ability to create what William Gibson called, in his book *Neuromancer*, a "consensual hallucination"—a group exercise in world creation where reality is no longer ordained from above, but generated by its participants. Fully evolved video game play, then, is total immersion in a world from within a participant's point of view, where the world itself reflects the values and actions of the player and his community members. Hierarchy is replaced with a weightless working out of largely unconscious preoccupations.

Dueling Joysticks

Thanks to their experience with video games, kids have a fundamentally different appreciation of the television image than their parents. They know it's up for grabs. While their parents sit in the living room passively absorbing network programming, the kids are down in the playroom zapping the Sega aliens on their own TV screen. The parents' underlying appetite is for easy entertainment or, at best, prepackaged information. Meanwhile, they bemoan the fact that their kids don't have attention spans long enough to endure such programming. The kids, on the other hand, rather than simply receiving media, are actively changing the image on the screen. Their television picture is not piped down into the home from some higher authority—it is an image that can be changed. When Dan Rather or Tom Brokaw shows up on the evening news, the screenager doesn't experience this broadcast as the gospel truth. To him, it's just a middle-aged man playing with *his* joystick.

The screenager sees how the entire mediaspace is a cooperative dream, made up of the combined projections of everyone who takes part. Today, Dan Rather and Geraldo Rivera have a bit more of a say in what that dream looks like than most of us, but even this is changing as more people get online and begin uploading their own text, images, and video. The difference between the Sega kid and Dan Rather is that while the non-networked video gamer is involved in an essentially masturbatory act, Rather is communicating and interacting with other people. Still, like any masturbator, the Sega kid is learning how to use his equipment. He orients himself to the television screen as a race car driver to his windshield. The games he plays are simulated drives through the very real data networks he will access later on with his computer and modem.

Market research indicates that screenagers are migrating from their game machines to legitimate personal computers in droves. "I stopped wasting my time on video games and started wasting it on the Internet," one appropriately cynical fourteen-year-old boy told *Upside*, a technology industry magazine. According to its article about the possible decline of video game consoles, a marketing survey conducted in 1995 indicated that kids overwhelmingly prefer PCs to game consoles as a platform for entertainment.* Further, psychologists have noticed an improvement in intellectual abilities—problem-solving, creativity, visual and spatial conceptualization—among college students who play video games regularly.† Researchers at New York University Medical Center use video games to improve hand-eye coordination in recovering stroke victims, and most computer training specialists use video games as a teaching tool for adults and children alike. If nothing else, they serve as excellent flight simulators for cyberspace.

But the promise of cyberspatial orientation is of little consolation to worried parents. The endless hype and panic over cyberspace, cybersex, or anything with the prefix *cyber* is a bit tiresome, but only natural. We are beginning to participate directly in something we've until now experienced only as programming from above. Imagine if you were told you could choose exactly what you were going to dream. The promise of cyberspace is the same. This is why, back in 1984, William Gibson imagined the "Net" as a consensual hallucination. Our media is our shared and weightless collective psyche. No wonder it evolves toward chaos.

*Richard Brandt, "Nintendo Battles for Its Life," *Upside* 7, no. 10 (October 1995).
†Patricia M. Greenfield, UCLA psychology professor, in *U.S. News and World Report*, 16 July 1990.

The Internet was originally designed for the use of scientists and researchers, most of whom had some connection with the defense industry. Although most summaries of the Internet's beginnings overemphasize its military component, the Defense Department did contract the think tank Rand Corporation to come up with plans for a nationwide computer network (Arpanet) that could function even after a significant portion of it had been destroyed in a nuclear war. The resulting network of computers was modeled on an organic structure rather than a hierarchical one. By decentralizing the network, its planners made it impervious to a pinpointed attack. Furthermore, the system was designed to repair itself by finding new routes for information when a particular path was down. Like a chain-link fence that can conduct electricity even if a hole is punched through a portion of it, the network can reroute messages around computers or lines that have gotten knocked out of service.

The kudzu-like sprawl of computers enmeshed thousands of institutions. Scientists posted laboratory findings and shared information through "newsgroups," which were organized by discipline or institution. The social aspect of the networked proved irresistible, and soon the scientists and researchers were sharing more about their private interests, hobbies, and favorite TV shows than they were about their research. The whole network was eventually retired from military use, but continued as the Internet, kept alive by the institutions maintaining host computers. Anyone with a computer and a modem can get access to the Internet in most areas for a small monthly fee.

The realm of computer networks is a created world, built upon an intentionally organic, anarchy-inspiring skeleton. No computer means more than any other, and only the weightless can pass between them. When the Internet's practical application gave way to pure pleasure, participants created bulletin boards and conferences dedicated to extremely personal, intensely spiritual, and highly philosophical subjects. It was as if going online somehow opened a person to a more chaotic sensibility. Fifty-year-old businessmen got into conversations about Carl Jung with teenagers whose dreadlocks and piercings would repulse them in the street. New kinds of forums arose to give people a chance to interact in more dreamlike ways. Multi-User Dimensions (MUDs) allowed users to engage in text-based fantasy games with strangers from around the world. Physically, it was as safe as sending e-mail; psychologically, well, it was as depraved as the participants wanted it to get.

At the moment we realize that the computer medium is not just for reading and consuming, but for posting and participating, and an entirely new set of responsibilities confronts us: What do we want to say and do, and what effect will our words and actions have on the consensual hallucination?

With increasing rapidity, the hierarchically structured databases, newsgroups, and file-sharing systems (FTP, Usenet, gopher servers) are giving way to more freeform-style Internet browsers, that encourage users to chart their own, almost random paths through the world of computer networks. The World Wide Web lets people participate online in a manner much more consistent with the underlying network. Any person or institution can create a "home page"—a bunch of data, images, text, and "links" to other Web pages throughout the network. Because these pages are linked to one another, users roam from one place to another, exploring the Web in the manner they would explore a natural environment: go to a tree, inspect its bark, see a bug, follow it to its nest. On the Web, one might start by accessing a computer in New York with a page someone created about Marshall McLuhan. He can click on a picture of the book *Understanding Media* and get connected to another Web page containing the text of the book. On that page, he may find links to other media theorists, television museums, or the University of Toronto. A click of the mouse takes you to the new site.

Exploring the Web requires a surfer's attitude toward data and ideas. Maybe this is why it took a kid to develop the software most people now use to navigate it. Back in 1993 at the age of twenty-two, Marc Andreessen, an avid PC game player, co-wrote the first version of Mosaic, the first widely used and user-friendly graphical interface for the World Wide Web. His stated aim has always been to create software that allows people to get the most out of existing technologies. Currently called Netscape, the later version of Andreessen's program allows users to access sound, graphics, and video over their phone lines, when previously only expensive phone connections were capable of such dense data transfer. No one thought it was possible, so nobody tried to do it. True to the screenage ethic, Andreessen waxes organismic when he talks about the Web: "The Internet is simply a communications medium," he told the screenager 'zine *Blaster* in one of his first major interviews. "It reflects society. . . . The fundamental nature of being human is always reflected in the communications medium. But the Internet will change how people communicate, how they conceive of themselves as a part of the world."*

Young people like Andreessen are at the forefront of networking technology. Even those who aren't programming wunderkinds are responsible for a great number of the most interesting Web sites around. This is because they aren't afraid of the seemingly nebulous quality of cyberspace, or of reevaluating how they are "part of the world." They have learned to get their bearings on the Web by anchoring themselves in the associations called links. The connective fibers from one site to another are the substance of cyberspace. As a result, there are two qualities that can make a Web site excit-

*Doug Millison, "Interview with Marc Andreessen," *Blaster* 1, no. 1 (January 1995).

ing and popular. It can have a great interface and data of its own so that visits to the site are fun and fruitful; or it can simply have great links on it. Some sites are almost entirely dedicated to giving people lists of other interesting places to go. To a screenager, this is information in itself.

McLuhan once said that the book was going to devolve into the blurb—the little quotes on the back cover from other authors or famous people. He meant that what is within any particular book will be less important than what someone said about it, and who said it. Well, now the blurb is (d)evolving into the Web link. A home page's character and strength—its ability to influence the rest of the Web—is wholly determined by how many other pages are linked to it.

Fruitopic Leverage

Hardly immune to the growing chaotic urge, the traditional media is slowly emerging from its own terrible two's into the more adolescent drive toward real participation. Camcorder tapes and public access television have changed the way television looks and behaves.

We have come to trust the grainy, bumpy footage of Rodney King and Bosnia over the high-gloss finish of staged press events. Always on the prowl for a new cult hit, and receptive to programming created by people just like us, we search out weird little programs that express highly personalized visions. A teenager named Jake produces *Squirt TV* out of his bedroom with a camcorder and sits on the edge of his bed irreverently chiding a rock star like Michael Stipe for shaving his head for his appearance on the MTV Awards: "Like we didn't know you were going bald, anyway!" Activists from around the nation gather their own camcorder footage and download satellite feed for compilation into alternative news shows like *Deep Dish TV*.

Mainstream programs and commercials now imitate the style of this guerrilla programming in the hope of attracting viewers and adding a sense of credibility. *ER*, *NYPD Blue*, AT&T commercials, and many other high-budget productions intentionally use jump cuts, hand-held cameras, and offbeat hosts to make their productions look viewer-generated rather than planned in lofty corporate boardrooms. Snapple's home-spun media campaign features "real people" who write in about their lives to a sweetly plump spokeswoman, who is herself shot with way too much headroom, much in the style of amateur home video. America's mega-beer manufacturers have launched fake microbrewery labels like Red Dog, as if the products and campaigns were the efforts of tiny, hip com-panies—remote high leverage points in a chaotic system. The marketers know that if they look like the "suits" they will fail in screenager-dominated mediaspace that no longer favors or respects the heavyweight.

Meanwhile, the programmers and advertisers who buck the trend fail miserably. Hollywood is still desperately trying to come up with movies and TV shows that depict the more frightening and dangerous aspects of our consensually hallu-cinatory mediaspace. None of the films or programs imitating the cyber experience succeed because they pretend that these technologies aren't readily available. In the movies, no one gets to travel through virtual reality without a room filled with expensive, unavailable computer equipment, and cyberspace adventures are far from removed from the reality of a real computer experience.

Fox TV's short-lived series *VR5* starred blond and beautiful Lori Singer as dreamy, antisocial telephone technician with mysterious psychic abilities that allowed her to shanghai unsuspecting people on the other end of her phone line into her own virtual reality nightmares. Her technical and super-natural abilities were so unique, in fact, that a mysterious conspiratorial brotherhood enlisted her in their efforts to, well, do some really nasty things. We were to look on, amazed, as she wandered through imaginary landscapes attempting to extract information about covert operations from the brotherhood's enemies. If this show were your only exposure to computer networks, you'd have to conclude that exploring virtual reality—if you ever could—is about as dangerous as *Star Trek* and paranoid as *The X-Files*.

John Landis produced another TV cyberthriller, the outlandish interdimensional travelogue *Sliders*. Here a genius college physics student from San Francisco develops a remote control device that allows him and his band of unlikelies to roam to and from alternative Earths. Each show takes place on an Earth both similar to and violently different from ours. Worst of all for this gang of unwilling cybernauts, neither the student nor his Nobel Prize–winning professor can figure out how to get back home to their own Earth. If only they could find the "escape" key. They should have just stayed at home and watched TV like all good Americans.

The parade of cyberthriller movies, including *Virtuosity*, *The Net*, and *Johnny Mnemonic* fail for similar reasons: They exchange the real thrill of participating on-line for the Hollywood theatrics of car chases, machine guns, and explosions. Going on-line may be dangerous to one's worldview, but not to one's physical being.

Why are the film and television industry intent on misrepresenting the chaotic quality of mediaspace? Because—all conspiracies aside—the long-term aims of big money moviemakers and TV producers are directly opposed to the possibilities opened by our new technologies. The movies need to make things look bigger than life, romantic, and far removed from the day-to-day reality of their audiences. That's why we pay to see them—because the worlds they open to us are inaccessible by any other means. They must be offered to us from above. So a TV version of virtual reality will never look like something you could rig up in your own bedroom with an appliance you bought at Circuit City, nor will the cinematic depiction of the Internet ever offer you the opportunity to type in a few keystrokes and find yourself in Tokyo.

Our established media outlets may be losing ground in the battle against cooperative, participatory media, but they are not going down without a fight. It's as if the traditional mainstream media—the parental expression of communications technology—*wants* us to be afraid of participating so that we never leave the safety of its nest. The producers and editors, steadfast in their belief that they can monopolize our airwaves and printing presses, probably have something to do with it, too.

Our newspapers tell us we are too stupid to go online and find information for ourselves. A *New York Times* piece on health and the Internet warned readers of the potentially damaging information available online: "Some of it is harmless, but a lot is dangerous to your health and well-being." The writer learned that, although one health forum recommended blue green algae as a dietary supplement and appetite suppressant, "their product was nothing more than pond scum."* That algae is pond scum is no more relevant to the argument of its efficacy than the fact that penicillin is nothing more than bread mold. The *Times* and medical doctors interviewed look down upon the homeopathic, chiropractic, and herbal treatments extolled online because they are "folkloric" in origin.

Folklore is an unthreatening mode of communication when it's applied to quaint cultural relics like flamenco dance steps or quilting patterns. Such hand-me-down cultural legacies had all but vanished in modern America. Now that the World Wide Web has restored grassroots communication, we are witnessing a reemergence of folkloric technologies. Online participants wander the global mediaspace like gypsies,

*Marian Burros, "Eating Well," *New York Times*, 1 February 1995.

gathering and sharing information as individuals, unassociated with institutional authority (that's why gypsies were so persecuted through history). The difference between folklore and AMA/FDA-approved medical research is that the end user is responsible for judging the folklore's effectiveness. On-line discussions of medicine treat health care like a Zagat's guide rates restaurants: Word of mouth and personal experience mean everything.

But medical professionals feel that we, the untrained laypeople, have no business circumventing the doctor and pharmacist, nor judging for ourselves how to treat our ailments. If the *Times* and our authorities had a leg to stand on in their efforts to deny us access to medical information, it would be to argue that treatment via Internet could lead us to recklessly self-medicate, avoid professional attention when we need it, go against our doctor's good judgment, or try bogus treatments while postponing effective ones. So far, though, that's not what's happening. Most people getting medical information on-line are suffering from mild symptoms and would have simply bought an over-the-counter remedy or just sat and suffered. In fact, it's often only after the insistence of a few on-line friends that someone seeks professional medical help at all. Others have already gone to the doctor and want a second opinion, or have been told that nothing in the traditional medical arsenal can cure them and are now turning to alternative methods like herbology or homeopathy. Still others, like cancer and AIDS patients, use the computer to learn about experimental therapies or to form support networks and "buyers' clubs" for new European medications unavailable by prescription in the United States.

Yet to go on-line and seek out our own medical answers is one of the most radically independent actions we can take in the modern world. Doctors may be the last authority figures we have left, and seem to hold the keys to an extensive and secret body of knowledge. As Internet-dystopian Clifford

Stoll pines in his book *Silicon Snake Oil*, the problem with using our computers to unlock this vault is that "it's up to the reader to separate out the dregs."* To explore this knowledge ourselves, taking responsibility for sifting the valuable from the nonsensical from the potentially dangerous, is to commit to the proposition that, armed with a laptop and a modem, we can explore any other technology we want to— even medicine.

Even if going on-line doesn't compromise our physical safety, though, our traditional media wants us to worry about catching a computer virus that could potentially debilitate our personal computer or office system. If we're supposed to believe the TV and newspaper reports, they pose a threat to everything from our home PCs to the American way. It's a classic mother's argument against intimacy: You may catch something.

The specific harbingers of viral doomsday, it turns out, are the very companies in the business of protecting us from infection. They send out alarming press releases describing dangerous new viruses about to sweep the world's computers, and then the media, anxious to discourage us from getting our news from alternative on-line sources, reprint the releases verbatim. The high-priced computer security firms generate more business for themselves while the media generates more viewer paranoia about turning off the TV and going on-line. The scam is further promoted by software manufacturers who, in an effort to stop illegal copying of their programs, perpetuate the notion that any disk that is not factory sealed will spread cyber-STDs.

By marketing paranoia, the people who hope to make a fast buck off computer security are actually making the on-line world a more dangerous place. Each new expensive antiviral program or security measure is a gauntlet thrown

---

*Clifford Stoll, *Silicon Snake Oil*.

down to young hackers everywhere, who then toil to crack the new gates. It's like fighting bacteria with antibiotics. There's just too many of them, mutating too fast. It's the hackers—the kids—who realize that the real damage inflicted by viral paranoia is on the way we think about our computers. In an age of AIDS, violence, and cultural isolation, our public computer bulletin boards are welcome respite from disease, terrorism, and prejudice. By instilling us with fear about what we may "catch" by sharing data with strangers, the overzealous advocates of computer security are like a cancer that kills its own host. Unwittingly, they serve as agents for our own fear of intimacy by giving us more excuses not to jump in.

Of course the best excuse to stay out of cyberspace, and even more important, keep your children out, is the media-sensationalized cyberporn community. In spite of its frightening cover depicting a horrified child at a computer terminal, *Time* magazine's coverage of the cyberporn phenomenon was almost balanced in comparison to most mainstream media's overblown stories. While the opening of the piece touted the frightening statistic that 83.5% of the digitized pictures stored in Usenet newsgroups are pornographic, buried later in the piece was the equally important statistic that these images represent only 3% of the total messages contained in Usenet groups, and that Usenet itself accounts for only 11.5% of all Internet traffic. So it's not really 83.5%, but 3% of 11.5%, or about one-third of 1% of Internet traffic that's devoted to pornographic images. Much less frightening. Further, almost all the images exchanged over the Internet have been scanned in from magazines already available in adult bookstores. The vast majority of cyberporn is exchanged over private adult bulletin boards that require a driver's license as proof of age.*

While the threat of a child molester procuring new underage victims through on-line services is not to be shrugged off, the incidence of such atrocities has certainly been overemphasized

*"Cyberporn," *Time*, 3 July 1995.

by the mainstream media. The danger is really no different from a kind-looking stranger inviting your child into his car, and probably less likely to result in rape. Of the 800,000 children reported missing in the United States, only about a dozen have been attributed to seductions conducted on-line luring children to some real location, and these cases were all highly publicized. Of that dozen, many turn out not to be the work of adult molesters at all. In one famous case about a gay fifteen-year-old boy from Washington who was sent a bus ticket by a man he met on America Online, the abductor turned out to be another gay minor.*

Ironically, the Internet is proving an ideal stalking ground not for sexual molesters and child pornographers, but for the FBI agents trying to catch them. On-line investigations of cyberporn rings require almost no legwork. Since on-line interactions are all traceable, the perpetrator is always just an e-mail address away. In one successful "raid," agents simply posed as would-be buyers and sexually willing minors. The rest was easy. Undercover cops posing as Catholic school girls on the New York subways hoping to snare flashers and rapists should only be so lucky.

Our legislators offer deeper insight to our real fear of the sexual nature of cyberspace. One bill by Senator Orrin Hatch expanded the definition of child pornography to include not just pictures of children having sex, but pictures in which children are made to appear that they are having sex. As Hatch told the *New York Times*, "Today, visual depictions of children engaging in any imaginable form of sexual conduct can be produced entirely by computer, without using children, thereby placing such depictions outside the scope of federal law."† By this logic, we should censor *Lolita* again for depicting a fictional but illegal encounter. (Or better, prose-

New York Press, 6 September 1995, and *Seattle Times*.
†Use of Computer Network for Child Sex Sets Off Raids," *New York Times*, 14 September 1995, p. 1.

cute the editors of *Time* magazine for their titillating cover photo—a "digital illustration" of a young child reacting to cyberpornographic images.) It's the limits of our imagination that are on trial here—not the evils of technology. Frankly, I'd prefer that pornographers create dirty computer-generated cartoons than eye my daughter as possible camera fodder.

Because of its freeform weightlessness, cyberspace feels like a place where we are likely to express the worst sides of ourselves. There are no rules and regulations. As Cokie Roberts described Internet users on *This Week with David Brinkley*, "It's as if their mothers disappeared and now they're allowed to say just awful things."* What do we do when Mommy and Daddy are out of the room? Masturbate ourselves into oblivion like monkeys wired up to an orgasm button? Kill children? Blow ourselves up? These options may sound silly now, but these are our fears as projected back to us by our media. Our television and print news are dedicated to making cyberspace into the instigator of the most dangerous and deviant behavior imaginable, from child pornography to the Oklahoma bombing, and worse. The underlying and paralyzing sentiment is that we must be protected from ourselves.

Screenage Politics:
Breaking Co-Dependency

We are afraid of the universal wash of our media ocean because, unlike our children, we can't recognize the bigger patterns in its overall structure. It seems random, purposeless, and devoid of morality. So we tend to depend, instead, on whatever buoys and other arbitrary points of reference we can impose on it. Like cartographers, we seek to understand

*Cokie Roberts on "This Week with David Brinkley," ABC, 28 January 1995.

the ocean as a grid of longitudes and latitudes, yet ignore the possibility for orienting ourselves, more like surfers, to its constantly changing ebbs and flows. Without clearly defined gradations and relationships, we don't trust our own impulses to play or work with one another without getting lost or losing control. But as the turbulence of the Screen Age breaks down the last of our flawed coping mechanisms, we find our kids developing new strategies for legislating a chaotic culture.

Until recently, duality has provided many convenient ways for guiding our decisions. Our legal, political, and economic systems, in particular, have benefited from the simplicity of breaking down complex dilemmas to yes or no, good or bad, guilty or innocent. As we all feel ourselves becoming part of a single but undefined global system, it likewise becomes more difficult to distinguish any one person, thing, or action from the rest. Did the Los Angeles riots occur because some cops beat an African American? Or because of socioeconomic conditions in South Central L.A.? Or because of the media? Or was it inspired by insurgents? As we develop the ability to recognize the patterns underlying the discontinuous onslaught of current events, how can we say exactly what led to what? How do we assess credit, and how do we assign blame? Even Charlie Manson begins to make sense when, in a 1980 jailhouse interview with Tom Snyder, he can answer the question "Were you responsible for the murder of Sharon Tate?" with a knowing "What is responsibility?"

As our media forced us to consider these issues, we began to look to the same media for some answers. The clear-cut black-and-white certainty of cop shows like *Dragnet* gave way to a myriad of lawyer shows, *The People's Court,* and cable's Court TV. We became more concerned with the pursuit of justice than with simple crime and punishment, and felt free to question the validity of formerly open-and-shut cases. This development has not been without its drawbacks. Our new and complex appreciation of the legal system has made it difficult for us to convict anyone of anything. Were the Menendez brothers really responsible for their actions, or were they the victims of child abuse? Were John DeLorean and Manuel Noriega entrapped by federal agents? Was the L.A. "Gang of Four" responsible for their beating of Reginald Denny, or did they merely succumb to crowd consciousness and years of oppression? Could O. J. Simpson *ever* have gotten a fair trial or was every U.S. citizen too personally involved in the case as a media spectator to be objective? Meanwhile, Jewish lawyers from the ACLU spend the institution's time and energy defending Nazis with whom they violently disagree. Something's got to give.

Our growing sense of interconnectedness fostered by media has all but paralyzed us. We are caught in a classically co-dependent relationship with one another, where we cannot convict anyone without, in some way, condemning ourselves in the process. In trying to apply Old Testament–style justice to a postlinear and hypermediated culture, we are exposing the inability of our dualistic models to adequately process a chaotic reality. And also, as in a co-dependent relationship, far too much of our motivation is fueled by guilt.

The maintenance of an "us and them" model of society draws artificial delineations between people. The distinctions between guilty and innocent, haves and have-nots, oppressors and oppressed, and even, arguably, blacks and whites are not organic but invented. While most of us would prefer to erase these distinctions, the methods we use to do so perpetuate the problem. Housing projects assuage the guilt of limousine liberals who prefer not to drive by slums on their way home to the suburbs, but these projects condemn their inhabitants to geographic isolation and disconnection from any natural urban community of their own. Entitlement programs, though initiated with the best of intentions, are a perfect way to disable the group being entitled.

The idea of a helping hand is nice at first glance, but when considered in an evolutionary context, it is revealed as anything but helping. Think for a moment: What would be the best way to keep minorities unable to compete in a competitive marketplace? Give them business without them having to learn how to get it. They'll achieve falsely inflated sales while the special standards are in place, but when the entitlements are dropped, they won't have the necessary skills to make it on their own. In the face of new efforts to dismantle entitlement programs, many of the policy's supporters argue that they offer the only way for women and minorities to become members of the "old boy network" of white men. This envy is precisely what perpetuates those networks. The institutions of exclusivity are crumbling under their own weight—joining them, or wanting to, adds to the illusion of their validity. When I asked Thabo Mbeke, a leader of the then newly instated African National Congress what he would do differently in creating a democratic South Africa than the United States did during the Civil Rights movement, he responded without hesitation, "No entitlement programs." Why? "Who is doing the entitling?" he replied.

Indeed, entitlement can serve to heighten the class system, breaking down a population into entitlers and entitlees. Like a class of children, the entitlees grow more dependent on and even more resentful toward the cultural institutions both feeding and repressing them. The only possible outcome, other than total complacency, is the angry rhetoric of Islamic extremists, who realize that the liberal agenda is just as, if not more, debilitating to their empowerment than the conservatives' more overtly heartless policies. The radical Islamics understand that the motivation behind most institutional charity is to maintain separation, assuage guilt, and stave off revolution.

The planet-embracing media forces us to reckon with our global interdependency. Television serves as a window to almost everywhere else in the world, making us conscious of what is happening to everyone else. Thanks to television, the bombs in Vietnam, Lebanon, and Bosnia are exploding in our living rooms, too. The faces of children starving in Bangladesh or Rwanda can not be ignored as complacently as mortality statistics in a United Nations census report. Far from making the world more illusory or its audience "desensitized," television makes the world beyond our own neighborhoods much more real—so real, in fact, that it becomes incumbent upon us to take action.

Maybe this is why, despite popular characterization of Generation X members as couch potatoes, the per capita rate of volunteerism and social service in the screenager generation is higher than it was during the height of the Peace Corps era in the late 1960s. The brand of activism they get involved in is different from that of their parents, too. A new breed of campus feminists, whose vision is espoused by twentysomething social theorists like Katie Roiphe, reject the protectionist stance of the traditional women's rights movement. They insist on taking equal responsibility for crimes like date rape, having concluded that labeling women as victims of sex rather than active participants contributes to their oppression. If alcohol is consumed on a date, we are warned by traditional date rape literature, the woman will be less equipped to fend off the man's advances, and the man less able to control his impulses. The supposition here, that the new feminists oppose, is that the man wants sex and the woman doesn't. Why must we see the man as "plying" the woman with drinks? The new feminists believe that crimes like date rape, as well as the day-later regret called morning-after blues, will be best averted by changing the fundamental assumptions we have made about sexuality.

One way to make women less victimized is to stop treating them as victims. Rather than instituting bizarre social codes at universities, where, during the sex act, partners must ask for specific permission at each stage of contact ("May I touch your breast now?"), the new generation of feminists aspires to open communication from the beginning. If women can admit to themselves and their partners when they *do* want sex, then the mating ritual will no longer require them to play the role of a defensive strategist being slowly defeated by the man's offensive passion. In other words, if the underlying dynamic of the sex act requires that women play the conquered, then date rape is a mere extension of the social norm. By creating new social regulations based on protecting the supposedly unwilling victims of sex, we exacerbate the false duality of dominant men and dominated women.

Most screenage political activism is geared at penetrating the awkward ineffectuality of existing social contracts. The policies in place don't work, in the twentysomething view, because they are at cross purposes to themselves. The old policies attempt to eradicate injustices by institutionalizing them and to encourage independence by infantilizing the oppressed. This is because the old policies conform to a nonorganic view of social structure. Their advocates work intentionally to stunt the natural progression of the cultural organism, assuming that without an artificially imposed set of disciplines, things would get out of control.

There are two main motivations for these sorts of policies: guilt and fear. The guilty seek to soothe their consciences for the injustices of the past by implementing policies they can see and feel in the present. Like the lover who ditched you and then calls "just to see how you're doing," the guilt-ridden policymaker is patronizing, controlling, and self-important. Those inspired to institute policy out of fear, on the other hand, have a better handle on the principles of feedback. They perceive the resentment of their underclass, and know that it will lead to social unrest unless at least token gestures are made at regular intervals. Again and, in this case intentionally, the policies function to maintain dualistic class and social constructs. In both varieties of policymaking, a naturally evolving culture is feared for its destructive capabilities. The guilt-stricken liberals fear the destructive power of the greedy ruling class while the class-enforcing conservatives fear the destructive capabilities of an irate oppressed class. Both liberals and conservatives can cite scores of examples from history of societies that have broken apart or committed atrocities when they failed to conform to an imposed political template. Hitler and American slavery top the list for dominator scenarios, while Lebanon, Rwanda, and South Central Los Angeles show the need for benevolent despots.

All of these examples, however, are not the result of lack of social programming, but failed social programming. They are not examples of unfettered human will aided by technology but of the deep repression of individuals and artificial polarization of populations. The success of Hitler's Nazis is often blamed on the ability of media to get out of control. In nearly every debate where I've argued for a freer media, my opponents eventually bring up Hitler as the dangerous end result of an unregulated mediaspace. Nothing could be further from the truth. If anything, his reign of terror was aided by his careful control of print and radio. The Allies, who didn't realize Hitler was using recording tape, were frequently baffled by his ability to broadcast his voice from different locations. A Hitler does not result from the ability of communities to network through technology in a free-form fashion. He results from the monopolization of technology and the instilling of fear and paranoia in a culture that must get its information from a single source rather than from one another. Participation in the media makes this monopolization impossible, but such participation is willfully self-denied by the fearful and the guilty. If we believe that the natural expression of a culture through its media is Nazi terrorism, then we will deny ourselves the ability to self-express and we will precipitate just such an outcome. No one is easier to control than a fearful, self-denying, and self-censoring population. Vigilance can be a very dangerous thing. We had better watch out for that.

Many Americans like to believe that the bloodshed in places like Rwanda and Los Angeles is a natural outcome of uncontrolled populations. To them, the black man represents something closer to nature—to the jungle. His behavior, especially in war zones like Los Angeles and African cities, demonstrates the unbridled expression of human nature without a society to protect him from himself. Not so. These failed societies are not indicative of a natural state, but rather of the failure of European imperialism to dominate an indigenous people and of American social engineering to integrate its own formerly captive population. Just because we throw up our hands in disgust and retreat from an African nation or an American inner city doesn't mean that the ensuing bloody war for control over the region isn't our fault. We are not witnessing the expression of natural chaos. We are observing postimperialist populations struggling toward self-governance, and welfare state–isolated containment zones looking internally for any self-organization they can find, even if it means street gangs. These people have been systematically, though in most cases not intentionally, trained and encouraged over time not to take care of themselves. This is because while we should have been promoting nature and cultural intimacy, we were busy fearing it.

Screenage-style politics encourage a more open approach toward governance. Instead of instituting policies in order to curb our natural behaviors—until now presumed to be selfish and evil—we are to erode such policies by creating an atmosphere of irreverent positivism. The screenage activist assumes a certain level of intelligence on the part of the general population and seeks to disseminate information, promote networking, and provoke action in as natural a way as possible. It amounts to a restoration of arrogant, youthful optimism.

Groups like Act-Up pinpointed the chief dangers of the AIDS crisis as ignorance and, worse, passivity. The entire strategy of this high-profile, media-conscious activist effort was to provoke a societal response to the AIDS virus. Members provided crucial information about the disease and launched persuasive and often shocking media viruses to gain publicity. These media pranks—such as interrupting network newscasts or conducting mock funerals outside the National Institutes of Health—all served to stimulate what Act-Up felt was an artificially stymied natural response to the disease.

This prankstering is fun. Whether it's neofeminist riot grrrls postering San Francisco with crudely satirical slogans or the GenX PAC "Lead or Leave" challenging congresspeople to sign a manifesto to serve the electorate, the new style of social action has a spark of life to it. It's fun to see Madonna dressed in nothing but an American flag urging MTV viewers to "Rock the Vote." There's something—dare we say—*youthful* about this media-conscious activism, no matter how old the people performing it. But hand in hand with its childlike exuberance is a demonstrated willingness to treat people like thinking adults who can handle the facts of life. Screenagers believe that the individuals making up our world must be trusted with the information and tools to analyze it.

Thanks to media and a willingness to use it, real people are going to gain the ability to influence the direction of our body politique. Our only choice is whether to arm them with the skills necessary to effect successful stewardship.

The Paranoid Reciprocal

The accommodation and integration of the unfettered, natural will into public affairs is not a *que sera, sera* social scheme. It may accelerate a few things, and it may be fun for the anarchists, but more than anything, it will nurture the development of a networked global culture/world consciousness. Many great thinkers fear such a development because they see it as a reversal of human evolution back to the stage of the herd. Colonial organisms are great for plankton, but human beings are different: superior. If we are indeed superior, it is because we have the ability to be conscious of what we do. Plankton and minnows do not understand that they are part of reefs and schools. They simply behave. If human beings move into a more organismic relationship with one another, it will be by choice. We are not bees. We will participate consciously.

This is why technology and media play such a crucial role in this next phase of our evolution, and why it is so important that we continue to raise children who are less afraid of our inventions than we are. Technology is the method by which we consciously rig the communicative fibers of our planetary brain. Whether we are engineering the genes of our offspring or simply choosing to enable call-waiting on our second phone line, we are at least indirectly participating in our own and one another's forward evolution. Even though it takes on a life of its own at some point, technology still feels like our own creation. We develop it willfully, which is why we feel it is incumbent upon us now to evaluate its purpose and efficacy in serving humanity and nature's goals. Caution and despair dominate most considerations of technology, usually motivated by a deeprooted fear of intimacy and a denial of natural evolution—themselves both based in a dualistic and paranoid worldview.

The brilliant if pessimistic social theorist Marshall McLuhan foresaw much of our media revolution, but always felt the need to associate technological progress with biological or cultural decay. He pointed out how rock music made us deaf, and television damaged our eyes. Every technological innovation, according to McLuhan, has a reciprocal effect negating it. This way, everything stays the same or, more likely, slowly gets worse. "A speed-up in communications," he warns, "always enables a central authority to extend its operations to more distant margins." True, but, as any hacker knows, it also enables those distant margins to extend *their* operations back to the central authority.

McLuhan understood that our world was being hard-wired together, but he had little faith in the ability of the individuals being wired up to make use of the expressive capacity of the technologies with which they were becoming more intimate. To McLuhanites the telephone is indeed an extension of the human nervous system—but an unwelcome one. We are signing a pact with the devil, and for the privilege of being able to ring up our friends, we become enslaved to a bell that might ring us at any moment. Each time the phone chimes, we have an automatic nervous response. (Imagine what your dog thinks when he sees you grunt, get up, and answer the phone.) We have welcomed technology into our otherwise organic selves, and will pay the price. We are the victims of our technology—or so the McLuhanites would have us believe.

Imagine, for a moment, a person sitting at his place of work, at the keyboard with a headset, microphone, virtual reality goggles, and, let's say, foot pedals. Do you see this person as victimized or empowered? Most socialists from McLuhan's school would see our thought experiment's test subject as an exploited worker. Management, in its endless hunger for higher productivity, has enslaved one of the innocent proletariat in a web of wires and electrodes. Every possible sensory organ of this poor slave has been physically violated and condemned to hard labor, in a scene even more horrendous than a sweatshop.

To a screenager, though, this individual could as easily be an empowered cybernaut, for whom work and play are indistinguishible aspects of life. He may have won his VR goggles through hard work and perseverance, in spite of his boss's technophobia. He may even be working at home, as the designer of a dazzling new CD-ROM game based on his own analysis of Jungian dream archetypes. He may even be fourteen years old. If he experiences himself as a creative participant in his employment rather than its helpless victim, then every tool at his disposal is another avenue for extending his mastery. His nervous system—his very awareness—expands with every new implement he acquires.

Victims see progress as a continuation of their own progression toward further victimization. They seek stasis, because at least their predicament will not get worse if it stays the same. They spend their energies doing what is necessary to stave off disaster. Empowered screenagers, on the other hand, see progress as an augmentation of their own journey toward empowerment and expansion. They spend their energies doing what they enjoy in order to get to do it some more. (This is what "to slack" really means.) Their jobs get easier and more fun when they've got better tools. They understand that enjoying work isn't a crime, and that turning creativity into a commodity is like getting paid for playing. But they have also necessarily dropped the duality implicit in the word *employment*. They are no longer passively being employed by an external force or person. Everyone from a store clerk or software designer to a garbage collector or medical doctor is an active participant in the development of culture or maintenance of life.

But some jobs just aren't fun. Still, screenage movies like *Clerks* and TV shows like *Friends* glorify the apparently menial tasks of clerking at a video store or serving coffee. It's all in the attitude and relationships. Data entry and code crunching—labors requiring more hands and eyes every day—are hardly as creative as designing video games or creating network software. The young people involved in these seemingly rote tasks have found ways to make them more interesting. The best companies employing screenagers welcome feedback and ideas from their workers about making tasks more efficient and products less expensive. Young workers with great ideas quickly move up in the ranks to become programmers and creative executives. Meanwhile the kids working at terminals use company networks and the Internet to maintain chat windows and discussions with one another during work hours.

Screenagers under the most repressive work conditions express their creativity to one another by hiding jokes, graphics, and sound files within the code of their programs. In one famous incident, a "nude" drawing was hidden in a video disk of the movie *Who Framed Roger Rabbit?* Even in a computer program that I produced, the code crunchers hid a sound bite from an old science fiction movie, which could only be accessed by kids who knew where and how to type in a secret password. The Macintosh system once housed a secret game that could only be played by typing a special command in the Notebook desk accessory and then dragging it onto the desktop. The codes for these built-in extras are exchanged over the Internet and constitute a fairly extensive underground software distribution network.

The mostly screenage online and HTML programmers, whether working for Amnesty International or General Motors, experience their work designing Web environments as entirely creative. Even if they are just developing marketing venues for already established companies, the programmers are changing the roles of consumers, giving them more information about products and more opportunity to feed back their own concerns and desires. The programmers are designing the future interface between people and their institutions and infecting their electronic venues with the screenager worldview. By focusing on the ability of technology to empower the formerly passive, young programmers are changing the face of commerce, and fostering a much more responsive set of industries and organizations.

Of the "real world" jobs that aren't fun, many could easily be done by computers or machines. But labor unions, with their members' best interests at heart, fight to keep the warm bodies in their income-generating if obsolete positions. Retraining these workers in post-Industrial Age skills would be the best option, but this would require a very new attitude toward and from the work force. As a member of the Writers Union, I receive monthly newsletters from its parent organization the United Auto Workers, in which all technological and social developments, from robotics to free trade, are framed as detrimental to the cause of the worker. Clearly, those unoriented to the screenager's skills and, more important, his attitude that work and play are indistinguishable are at a disadvantage.

But most adults are afraid of play and suspicious of fun. They believe only kids enjoy themselves, and that they do so in blissful ignorance. How could a job ever be as much fun as *Mortal Kombat II*? And why let our kids play these video games if it's only to train them to be computer workers in techno-sweatshops run by the Japanese on American soil? The blurring of the line between work and play breeds paranoia. People are scared that their play time is actually camouflaged work. I've watched a number of different people respond to the AT&T television campaign where people are shown utilizing futuristic technology. A man sends a fax from the beach, or a woman calls home from a wrist-phone, and an announcer explains how "you will" be doing this someday. The common response to the ads is: "What if I don't want to?" The bogeyman here is that our leisure time is about to be gobbled up by our employers or clients, wherever we might be vacationing. We don't buy the notion that we could be spending a workday on the beach or in the mountains instead of in an office—only that our free time in the mountains could be interrupted by our work responsibilities. To adults, the commercials portend "the end of free time as we know it." No matter how natural and fun it may look, when the insistence to partake of technology comes from Ma Bell, it feels like an imposition. When it arises naturally out of the desires of individuals, it can serve a very different function.

We still don't trust the institutions that have appeared to

dominate our lifestyles to this day, and we can't envision a way for technology to liberate us from their control. McLuhan understood that organized, hierarchical systems lead to discontent, instability, and centralized power, while natural, chaotic systems promote the reverse. He liked the "mild institutionalization" of the village, because "everyone could play many roles. Participation was high, and organization was low. This is the formula for stability in any type of organization."* He laments the growth of the village into the city-state, which forced people to specialize and sacrifice their overall participation in local affairs. Similarly, he believed that technology forces further specialization and even makes simpler, craft-oriented specialties obsolete.

Up to a point. What McLuhan failed to foresee was that technology and cities alike would become so complex that their linear, highly organized structure would give way. The World Wide Web is anything if not highly participatory and barely organized. Computers don't "oversimplify" human interaction, as leading technology critics like Jerry Manders argue, they make it much more complex—almost organic. The tremendous stability of the current vanguard of computer community-making is the result of the rather random way it sprang up. Whoever understood how to create a site and had access to the proper equipment could do so. Like any people exploring a new frontier, they formed villages, alliances, and networks for mutual support. Eventually businesses established presences on the Web, encouraging people to visit their sites and learn about their goods and services. Many of these businesses are Web-related, offering software or advice for creating Web sites or navigating the Internet. And again, as most urban planners now recognize, the most stable and thriving communities are built around business and commerce. They are not isolated suburbs, but small towns where

*Marshall McLuhan, *Understanding Media*, p. 97.

people can gather naturally around barber shops, corner stores, and banks where they conduct their daily business. On the World Wide Web, just as in the well-designed town square, work, entertainment, and the civil society are interdependent and ultimately indistinguishable.

But when a technophobe visits the World Wide Web he has one of two negative reactions. He is either overwhelmed by the chaos, lack of censorship, and absense of formal authority, or immediately suspicious of the corporate entities who are staking out their own turf on the net. The former reaction is akin to the fear that we don't know what's good for us. Left to our own devices, we will make a mess of things. Even if that's true, we would probably do that anyway, with or without a Web. The latter argument is a bit more insidious, because it is held by many of the cyberutopians who should be championing the Web's development. The fear assumes that once business appears in a communications network, the network is somehow polluted. A Web site will now be used to "make" a person buy products. What should, according to the utopians, be a free and open system with uncommodified information will turn into a commercial jungle.

This attitude kills economies. If we are to strive toward an information and idea-based culture, then data and creativity will become the new currency. What is the alternative? Fight against the commodification of information and creativity in order to hold on to our Machine Age union jobs? The same ones that McLuhan and others blame for culture-deadening specialization in the first place? Many still wonder how people are going to make a living in the so-called Information Age. We will do so by selling what we generate from our minds and on our own computers as text, image, and code. When the commodity we have been selling—our physical labor—becomes obsolete (or at least less in demand), there had better be a new commodity to sell. Try creativity. Infinite supply, environmentally safe, culturally valuable, and even fun to make. If all our information were destined to be "free," then no one would be able to make any money with it, and we all really would be in need of some gainful employment, fast. The robots are already making our automobiles. The reason we don't need to worry is that the creation of wealth without the exploitation of physical resources has become a reality. In order to take advantage of this opportunity, we must learn to see ourselves as masters rather than victims of our new communicative pathways.

For example, advertising and public relations may not really be so evil after all. They are simply the technologies for packaging of data in an information culture. Advertising is itself a key commodity. It codifies data into recognizable forms that get the attention of intended targets. Of course, at almost every talk I've given about networking technology, someone complains that the Web might be a new avenue for disinformation, especially by advertisers. How will a person know what to believe and what not to believe? Compounding the confusion, reporters sometimes research a story simply by visiting a company's Web site and transcribing prepared press materials. They quote the press releases nearly word for word in their news stories, leading those of us who have seen both the posted press release and the published article to become quite suspicious of the way news is being gathered in the Information Age. What we fail to realize is that this is the way most reporters have *always* researched their stories. Before computer networks, it was accomplished more surreptitiously, through mailed press releases or phone calls. Now that we are all becoming witnesses to the way information is disseminated, we are gaining a more, not less, accurate picture of the ways disinformation is accomplished. The process is being revealed to us.

Which in turn leads to more fear and paranoia. If everything is being revealed, what happens to privacy? The arguments against mediating technologies pendulum back and forth, from one nightmare scenario to another. The telephone is too intrusive? Get an answering machine. But doesn't an answering machine allow people to filter out those with whom they do not wish to speak, promoting isolation and the censorship of opposing views? Most people are still utterly confounded by the moral dilemma of call-waiting. If we let the new call through, we will be forced to decide who is the more important call and express this priority to both callers. Or lie as skillfully as possible.

The real problem is that the more convenient our communications technology gets, the more choice we have. The more choices we make, the more honest we get, and the fewer secrets we are allowed to keep.

This is why we are so threatened by the clairvoyant children of animé or the mind-reading mutant alien tots in John Carpenter's recent remake of *Village of the Damned*. If we become part of a mediated colonial organism, other people will know what we are thinking and feeling. Last year, I broke up with a girlfriend after telling her only half the reason why. She had access to my America Online account and used my password to review my e-mail to friends in which I had more honestly expressed my feelings. When she confronted me, I was both angry and deeply embarrassed. But in the end, I was simply forced to admit what I had really been feeling, which I should have done in the first place. No harm, no foul. It's time we get honest.

Weightless Politics

Like sex, participatory media allows for high levels of intimacy while generating great news copy about loss of control, fear of disease, and imperiled children. Back in the Middle Ages, the Church condemned its followers' raw sexual energy and rechanneled it into religious fervor and devotional activities. Old English poems about physical lust were rewritten to direct that passion toward Christ and Mary, and words meant to evoke the phallus were changed to signify the crucifix. Today we willingly cut ourselves off at the wrists from our keyboards, lest we lose the sense of direction that our moral guardians have supplied for so long. Unlike our kids, whose Sony Playstations teach them the joys of self-navigation, we yearn for guidance from above and restrictions on how and where we move.

As those guiding authorities lose the ability to convince us of their moral high ground and intellectual superiority, we need to adjust our social institutions to support the sustenance of the non-hierarchical, user-generated reality that is fast approaching. This means combating the fear of stupidity, vulnerability, innocence, and evil that we are currently fomenting in our media and learning how to instill confidence instead. We must become willing to take responsibility for the world we are dreaming up together, however frightening or preposterous this may seem. We are like our ancestors who, understanding gravity only as a top-down proposition, couldn't figure out why people "down" in Australia didn't fall off the planet. Similarly, just because regular ol' people like us are steering civilization doesn't mean reality itself is going to fall apart.

Teachers feel the impact of empowering technologies first. Computers challenge the teacher's role as the classroom's chief information provider. A single teacher's brain can't hold much more teachable data than a couple of CD-ROMS, if that. When a kid can log on to an information service and gather facts about almost any subject, at a depth beyond any single human being's capability to provide it to him, his teacher must stop seeing himself as the storehouse of knowledge. Teachers threatened by technology attempt to restrict it, or even prohibit its use in the classroom, justifying their actions with bogus claims about how computers quell creativity or stunt social skills. This tactic, aimed at prolonging a teacher's monopoly on data, is doomed to failure.

Instead, like movie theater owners reckoning with the advent of videocassettes, teachers must discover what they can offer that a computer cannot. Such teachers will realize that they have been liberated from the rote task of supplying information—a machine can do that. Unlike a computer, a human teacher can be a partner in learning and dedicate himself to giving his pupils the necessary criteria to judge their data's integrity, make connections between different facts, and formulate opinions and arguments of their own. The best teachers will instill in their pupils the confidence and enthusiasm to express themselves as widely and articulately as possible.

Religious leaders will also have to alter their perception of their own roles. It won't be the first time. The Gutenberg Bible changed the priest's relationship to his parish. Once people could read, they brought their Bibles home with them and reconsidered the interpretations offered by their ministers. Martin Luther and others disseminated their antiauthoritarian opinions about Catholicism using movable type, and Protestantism was born. Today's religious leaders, too, will be forced to make some major adjustments as people try to come to terms with the spiritual implications of the consensual hallucination. Like trusted guides on the spiritual journey, instead of narrators of a vicarious one, religious leaders will be valued for their good counsel rather than their access to a "direct line" to God.

As schools, churches, and other institutions lose what's left of their tenuous hold over society's organization, it's no wonder that politicians are scrambling to build new garrisons. Both the left and right sides of the political arena argue for institutionalized regulation of human behavior. No one trusts a culture left to its own devices. The only real difference between the parties is the scale of societal architecture they want to put in place.

These days, Democrats and liberals have cast themselves as the advocates of centralized authority. Building on the model of government as family, they feel that evenhanded, nonprejudicial regulation, legislated by caring lawmakers, will prevent the weak and impoverished from falling off the map altogether. Social programs implemented by the federal government would ensure that all citizens, no matter where they live, get a fair shake. This, after all, is the point of these programs: to enforce a family atmosphere on a national level. Businessmen and powerful individuals are not to be trusted but kept in check through carefully erected regulatory agencies. The best of Democrats attempt to establish a balance between fostering the natural growth of capitalism and answering the need for gentle parental supervision.

The conservative argument appears more pro-chaotic at first glance, which is why it is enjoying such popularity among futurists and neo-libertarians. Republicans try to reduce the role of the federal government in business and personal affairs. The less stopgaps on the economic engine, the more efficiently it will run, and the more money and freedom for us all. Republican senators make long, impassioned speeches about the "rights of the individual" and generally support the right to bear arms, the rights of landowners over regulating environmentalists, and the right to run your business pretty much how you want. By eliminating social programs and public assistance, they feel they can toughen up the downtrodden rather than cater to their laziness.

The nonrestrictive, antibureaucratic bent of the conservative agenda, though ultimately only superficial, explains why the ideology is enjoying such a resurgence today in the age of the Internet, personal expression, and organic understandings of culture. The fall of the Soviet Union served as formal proof that centralized bureacracy just doesn't work. While at first the end of the Cold War hurt conservatives, whose hard-line foreign policy felt less necessary without a Soviet threat, the failure of Communism dealt greater damage to the liberals, who now represented the global left. Communism mistrusted the individual and equated the profit motive and personal gain with decadence and evil. All human efforts in a Communist nation, by law, were to benefit the state. Once the Cold War degenerated into a financial holdout it was only a matter of time before the United States' more robust economy won the standoff. America's monetary advantage and Gorbachev's vision for a more open society won the West its Cold War and demonstrated the strength of free market capitalism over a federal bureaucracy.

Deep down, however, the conservative policies are no less personally restrictive and chaos-dampening than the liberal ones. The only real difference between them is on what scale the left or right chooses to draw the boundaries of institutional control. It's no coincidence that the less-restrictive economic policies of the right are coupled with the agendas of the Christian Coalition, Moral Majority, and advocates of "family values." Society can be absolutely free, as long as no one has an abortion, experiments with drugs, listens to rock music with dirty lyrics, or engages in homosexuality. Prayer in school and tougher drug laws—not sex education or rehabilitation programs—are to strengthen the moral fiber of our young people. The conservative strategy is to institutionalize the family and the local parish.

By arguing for moral codes and family values, the conservatives are using a version of Christianity to fight the chaos of an organic culture. Society is allowed to remain unfettered on the level of federal intervention, as long as we can be sure that everyone is behaving properly on a local or familial level. The family is to become like a tiny corporation: The father sits at the head of the table (like God intended him to) and administers the family's values to his employees. Mom is a majority shareholder and plays chief operating officer to Dad's CEO/chairman of the board. The unwieldy reality is a nation as large and diverse as the United States micromanaged on the level of an individual nuclear family. The resulting mosaic of tiny square tiles is no less restrictive than the liberals' giant Playschool blocks. Both visions are simply overlays of imposed order on a free-flowing fractal system. The only difference is on which hierarchical level of the fractal pattern to build the dams.

The myth of the conservative point of view is that we have somehow lost a sense of family values in this country. On the contrary, the family may be one of the only values we have left—at least in spirit. What we have lost is a sense of community values, and the family is being asked to pick up the slack. Urban planning, housing projects, purgatorial suburbs, and poor communication combined to dissolve the natural bonds of community within a nation of immigrants. We became family units, cut off from one another, each as sad and unfulfilled as our neighbors, but afraid to admit the truth.

The prosperity of the post–World War II baby-boom era, by decreasing the obvious survival necessity for community values, destroyed what was left of the natural social scheme. Family values were really just a marketing concept, designed to sell the highest volume of products to the richest people in the history of the world. How do we get every single family on the block to buy a product—like a barbecue grill—when just one nice one would do for all of them, and probably be more fun? Instill a sense of competition among families. *Be the first on your block!* Woefully, this was done at the direct expense of community values. To keep up with the Joneses, you have to see them as the enemy.

Those rich enough to do so rushed out to the suburbs in their station wagons; those who couldn't afford to get out were left behind in the fiscally depleted urban wastelands. With family values an accepted morality, this abandonment was easy to justify. *Screw 'em. I'm helping my family. I love them, and no one can tell me not to get the best for them.* Just don't look back at those cities. If you do, simply rationalize that their poverty is their own fault. *Besides . . . those city people don't have family values.*

But those who missed out on the hollow American Dream of a united nuclear family are developing their own community values, however distortedly. The poorest of fatherless children band together in street gangs with violent initiation rituals reminiscent of those employed by ancient tribal communities. Although the shared activities of most urban street gangs are drug trafficking or murder, their shared bond is real. This is why white kids in the suburbs have adopted the style and attitude of their poorer inner-city black counterparts. Their need for a sense of community and shared values is just as strong—and just as absent from their everyday cultural experience. Ironically, perhaps, they yearn for the camaraderie of the urban wasteland their parents fled. Neither black nor white kids necessarily want to be gangsters, but the illegally obtained guns and drugs, as well as the codified "colors" and language associated with street gangs create a sense of membership. Rebelling against one's own parents may create a feeling of personal identity, but rebelling against the cops creates one of community and cultural identity. Gang fashions are based on the denim and oversized shirts of an even more "exclusive" club, jail prisoners. Being institutionalized accredits gang members' roles as outcasts and renegades. They are prisoners of the overculture; their clothes, language, and music are the symbols and media of defiance and unity.

Much like the Internet subculture, rap and hip-hop culture were enabled by a willingness to exploit do-it-yourself technology. Using tape loops or simple digital recording equipment, rap musicians can "sample" and remix the riffs of their favorite artists and then create their own lyrics as overdubs.

The music evolves as new artists layer their own sounds and words over existing tracks. Each song amounts to a cultural record of everything that went before it. The rap lyrics themselves are codified, much like slave spirituals of the early nineteenth century, so that singers can exchange urban coping strategies without the oppressor's knowledge.

When the would-be censors do catch wind of what the kids are talking about, they are horrified. Time Warner is regularly in the headlines for distributing music that, at least on the surface, degrades women and challenges the authority of police officers. The real and much more powerful threat of this music, however, is that it fosters a tight-knit subculture of kids who are willfully reprogramming themselves with ideas they feel are more appropriate to survival in the modern urban landscape. The kids are creating tribes bound together by new sets of values that, because they are necessarily disconnected from the values of their elders, unfortunately often lack some of the tempered wisdom of an older civilization. But can we blame them? By rejecting their efforts at community-making wholesale, we drive them further away and isolate ourselves from their quite healthy urge toward restoring a social fabric. Neither we, nor our kids, can go it alone.

Poet and social theorist Robert Bly makes a strong case for the initiation of young men into a cultural continuum beyond the boundaries of his own family. In most cultures, a group of elder men from the community welcome a boy into manhood. This is how children grow up. Moreover, the culture is permitted to develop as new youths, each with his own ideas, are invited to participate as adults in the greater community. Without such an evolving community, we all stay children.

Only by abandoning the need for enforced social and economic hierarchy and division as well as the convenient barricades they offer, and trusting that a weightless world will develop naturally into real, if fluid, communities, can we move out from our self-imposed parental control into true adulthood. Will a fourteen-year-old kid playing Mortal Kombat over phone lines with another kid a thousand miles away, or another exchanging hip-hop tapes with a "homey" from across town eradicate the past fifty years of community disintegration? Probably not. But by focusing on the *experience* of real connection to one another—which has nothing to do with defining an up or down, yet everything to do with gaining one's bearings in an intrinsically weightless system— we can instill in ourselves the necessary confidence to step out of the womb and into the unknown.

5

———◆——◆——

THE FALL OF METAPHOR
AND THE RISE OF
RECAPITULATION

Lot's wife behind him looked back, and she became a
pillar of salt.

Genesis 19:26

Our cave-dwelling ancestors gathered around the fire each
evening and told stories to one another. From what we can
surmise today, these were not elaborate mythical or moral
tales, but simple stories relating real-life hunting adventures,
battles against the elements, and the true migrations of their
own ancestors. The stories had a practical, communicative
purpose. When one man learned that "playing dead" could
fool an angry bear into leaving him alone, the rest of his clan
would benefit from his experience by hearing him tell the
story that night. To be sure, the gathering had entertainment
value and the better storytellers probably got more intense

reactions from their peers, but the stories themselves were literal and their purpose was instructional. Their value as amusement was directly proportional to their connection to basic survival skills.

As we developed agriculture and specialization, our best storytellers became professionals, who wandered from town to town telling stories less about particular tasks or adventures and more about themes and issues of interest to everyone, whatever their job or station in life. These were the epic stories of emperors and soldiers, monsters and heroes, or gods and virgins. The stories worked on a metaphorical level. No, none of the audience had ever traveled hundreds of miles from Greece on a boat or fought a Cyclops, but everyone had experienced leaving a place of some sort and trying to get back, so they could relate to the *Odyssey* as parable. Theater, films, and eventually television worked the same way: they all tell a story about someone whose plight somehow symbolizes our own.

This kind of storytelling requires its audience to make a logical leap: "That character *is like* me." Once the identification has been made, the plight of the character becomes a metaphor for the plight of the audience. It's *as if* it were you. The hero's choices and the results of those choices can be used to make almost any point, and very persuasively. We want to learn from the hero's mistakes. But the storyteller can stack the deck any way he wants, because the circumstances are completely within his control. This is why metaphor and morality go hand in hand. The church used theater to tell morality tales because it worked. Lot's wife disobeyed God's instructions and looked back at burning Sodom. She got turned into a pillar of salt. Whoops ... better not disobey God's law or that will happen to me, too.

To some extent, every play, film, and television show is a story of Everyman. We get engrossed once we've made the connection between the protagonist's plight and our own—but we relinquish control of the character as well as the qualities of his universe to the playwright. It's not real. It's a linear story, and one with a point. Someone else's. And this is why our kids won't stand for it any longer.

Metaphorical stories work in a neat, linear world. They worked when we yearned for guidance from above. They worked before we discovered that each of us has his own stories to tell, and that one overgeneralized parable about a Mediterranean sailor or even a Middle Eastern carpenter could never reflect the multiplicity of our experience. The purpose of parable itself is to simplify and reduce. The *parabola* is the U-shaped curve from Euclidean geometry that represents the relationship between a single point and line. (Technically, it is the set of points equidistant from both.) Likewise, a parable attempts to relate a single point or moral to the continuum of our lives. One story applies to us all, as long as we stand in line.

Well, in the simple theoretical world of an X-Y plane, such a curve can be drawn. In order to explain the chaotic natural world, however, mathematicians have replaced the smooth-edged figures of Euclidean geometry with nonlinear equations and fractals. In reckoning with the complexity of the postmodern experience, we also need to replace the parable with a storytelling form more suitable to a real world based in chaos.

Our kids have already moved from a metaphorical appreciation of parable toward a much more self-conscious, recapitulated experience of storytelling. Instead of drawing an audience into the plight of the protagonist, recapitulated storytelling intentionally distances them from the emotional reality of the plot. As playwrights since Shakespeare have understood, the way to accomplish this is by staging plays within the action of the play. The play-within-the-play, like those performed by the acting troupes in *Hamlet* or *Midsummer Night's Dream*, just like a painting-within-a-painting, reminds the audience that they, too, are watching a play. Instead of suspending our disbelief, such self-similarity calls attention to our very real relationship to the action on stage. We are an audience, engaged in our own experience.

By the twentieth century, as painters like Picasso played more self-consciously with point of view, playwrights like Bertolt Brecht did the same thing with theater, intentionally bringing focus to the relationship of the audience to the stage. He developed a technique he called the "alienation effect," which highlighted the artifice of the theatrical event by revealing the actors changing costumes and never attempting to hide lighting equipment or set pieces. His belief was that an audience lost within the reality of a play can never be moved to action. They must stay alert and disengaged, constantly aware of the inability of moral platitudes—the metaphorical existence portrayed on stage—to answer complex human dilemmas. By watching plays-within-plays self-consciously performed on stage, we become part of self-similar series of levels on which the same issues are being recapitulated. Instead of looking within the context of the play on stage for the answers that Brecht's tragic characters could not find, we look outside the theater into our very real world.

Television since the 1970s has used similar forms of alienation in an effort to ask more questions than it answers. The "master shot" used to photograph *All in the Family*, for example, was from the point of view of Archie Bunker's television set. Just as Archie observed the world around him through his screen, we observed his through ours. Our TV sets served as portals into each other's living rooms. But we were aware of this relationship while Archie wasn't. He looked naïvely into his set for answers, while we watched, ironically and self-consciously, for comedy. Unlike a stage-play program like, say, *The Honeymooners*, the show was able to tackle some of the most relevant social issues of the real-world 1970s because it recapitulated our own viewing experience and kept us aware of our relationship to the screen.

By the 1990s we had come even further. The opening theme of today's politico-sitcom family, *The Simpsons*, is played over animation of the entire family rushing home to the living room couch in time for their favorite show—recapitulating, the producers hope, the rush to the TV set in homes across America as *The Simpsons* theme music begins. The show is satirical in nature. The distance and self-consciousness it provides gives us the room to experience irony. Mirroring our increasingly ironic sensibility, the program's child protagonist, Bart Simpson, seems aware of his own role within the show and often comments on what his family must look like to the audience watching along.

Kids today watch all television in a manner designed to promote this same self-consciously objective distance. *Melrose Place*, one of the most popular programs among high school and college students, is usually watched in groups who comment on the action almost continuously. This style of viewing is so popular that a bar in Los Angeles has been named Melrose Place and, like a sports bar, airs the program to large crowds of screaming, wisecracking fans. The action of the show is intentionally and outlandishly melodramatic, and the dialogue is delivered slowly, with long silences between lines so that viewers have ample opportunity to make jokes or predict characters' responses. The program itself is merely fodder for a mediated but witty conversation between friends. It is stimulus for a social event, even a Web site, and not an end in itself.

The television becomes a hearth, casting light on a room filled with participants. By making constant commentary, the members of the group watching prevent one another from becoming too involved in the action. Like the campfire of their ancestors, the television is merely a point of focus. But instead of just casting light on those telling the stories, the TV recapitulates the intrigues, affairs, and ironic disposition of its viewers, without demanding attention or engagement and never serving up moral templates for anyone else. (In fact, usually it's the villain who gets rewarded.) Shows like *Melrose Place* only make sense in the context of the social scene in which they are enjoyed, when they are personalized by the retorts of the kids watching.

Those of us unaccustomed to recapitulated media balk at our kids' viewing habits and accuse them of being unable to simply shut up and watch. We can't see that they're the ones who have developed the most advanced style of TV viewership: distanced participation. It allows them to turn what used to be an isolating experience into an expression of community. Like the fictional residents of the Melrose Place garden apartments, the kids have managed to create a social organism around a TV show. Maybe the great crime of the 1990s—irony—is not such an awful thing after all. We associate "self-consciousness" with a gawky teenager, hyper-aware of his own physical or social shortcomings. It seems, however, that a little self-consciousness can liberate us from the dehumanizing effects of reductionist and metaphorical storytelling.

All cultural developments follow this same three-stage process: an innocent literal stage, followed by a symbolic or metaphorical stage, and then finally an ironic recapitulated stage. These are not necessarily cultural epochs like social theorist Alvin Toffler's three "waves," but stages that can be observed in many different arenas, each occurring on their own schedules.

Computers and online networking, for example, began in the literal stage. Simple written commands in operating systems called UNIX and DOS expressed, literally, what we wanted to do. We typed "telnet" to log from one computer to another, or "dir" to see what files were in a particular directory on a disk or hard drive. To make this process easier for those of us who didn't know how computers work, commercial online services developed metaphors—like those used on Macintosh computers—of folders, files, desktops, and remote controls from the physical world to symbolize what was going on in the computer. Finally, as we get more comfortable with computers and as the services they offer us get more complex, we want more control over what the computers are doing. We want to be able to understand and relate directly to their functioning. The World Wide Web and the browsing software developed to navigate it are the first of the postmetaphorical, or recapitulated computer languages. Users log on to sites by clicking on their names and addresses. Downloading information is not literally ordered by typing a word command, or symbolized by dragging a "file" into a "folder," but recapitulated in the action of clicking on the very thing we want the computer to "get."

Money also passed through these same three stages. Gold was a currency with actual value. Gold and silver certificates represented precious metals, and then Federal Reserve notes recapitulated the original value of money, which was to "store" work or commodities as currency. For modern money and credit to work, we must be conscious of its value as a unit of exchange.

Our fictional television programs, though part of the larger process by which our storytelling developed, also passed through the three developmental phases. The first programs were simply broadcast stage plays—a direct translation of one medium to another. Eventually shows attempted to look absolutely real. We were to relate to the characters as if they were real people, in real situations. Finally, programs like *NYPD Blue* and *ER* emerged, whose reality was based in keeping us aware of the camera. But true recapitulated media is reserved for the kids who are ready. *Beavis and Butt-head*, though dangerously mindless in the view of most parents, artfully recapitulates the experience of kids watching MTV. As the two knuckleheads comment on the rock videos they watch, they keep audience members aware of their own relationship to MTV imagery.

Why is recapitulation necessarily more advanced or better than literal or metaphorical understandings of our world? Because it is capable of representing our chaotic cultural experience in a manner that allows us to relate to it. It gives us an insight into how nature works, and motivates us to become more fully conscious and self-determining. Unlike literal models, recapitulation doesn't demand that we memorize facts and commands, especially when there are too many to keep track of. Unlike metaphor, recapitulation doesn't demand a definite but potentially disastrous conclusion.

Besides—it's what the kids are doing.

Recapitulation *vs.* Gamara

We experience recapitulation before we are even born. Although scientists are still arguing about exactly what sort of evolutionary event takes place in the womb and what it means, it's pretty clear even to a layperson that ontogeny, to some extent, recapitulates phylogeny. That is, in the womb the developing fetus passes through the same stages of evolution that human beings did, from worm and fish through mouse and monkey. Little fetuses develop gills and tails before they become tiny humans. Embryonic development recapitulates evolutionary development. Modern scientists insist that the fetus is not really passing through true evolution, but merely passing through the embryonic stages of our evolutionary ancestors. Yet, in the womb a human fetus takes the form of a fish fetus, monkey fetus, and so on. This is not metaphorical. It is not *as if* the fetus experiences the history of evolution. It quite literally passes through each stage as it evolves. Our species' evolution is recapitulated in our personal evolution from ovum to baby. Adult creatures have no direct experience of this. It's funny that recapitulation, even on a biological level, is something experienced only by the young. Once we're born and individuated, environmental factors appear to have much more influence over our lives than our genetic history.

The other reason scientists are having trouble with the theory of embryonic recapitulation is that it doesn't exactly work out. The direct, linear timing of certain evolutionary features as they appear on the fetus are slightly out of order when compared with what we know about how evolution really occurred. But all this means is that the real-life execution of various triggers in the DNA coding didn't emerge exactly on schedule. No matter how well a picnic basket is packed, someone might still need something from the bottom before the top is unloaded. Our evolution isn't wholly predetermined by DNA, but neither is it just a random process of natural selection. It's chaotic, but there is a natural, underlying order to it.

Cultural ideas occur out of order, too. Copernicus and Giordano Bruno came up with true theories that were intolerably divergent from the conceptual order of their day. We simply weren't ready yet. Rather than accept ideas that would have cracked our reality, we just killed the voices espousing them.

To accept a recapitulated model of our world today, we must learn to reckon with the no-man's-land between predestiny and self-determination, or ordained order and total randomness. It means getting used to a world that looks and acts something like a fractal. Large motions are recapitulated by the tiniest ones in amazing but inexact self-similarity. The shapes magnified in the smallest details of a fractal are self-similar but not identical to the larger ones. They resemble each other enough for us to recognize them, but not so perfectly the same that we can pinpoint them as exactly congruent. Further complicating matters, just because they are occurring on different levels doesn't mean—as in metaphor—they are unrelated to each other. They are all part of the same system. If a certain species of microbe in a coral reef begins to perish, the character of the entire reef will change.

The trick to getting comfortable in a chaotic natural world is learning to recognize self-similar patterns, even though they don't exactly mirror one another. Like the computers we've programmed with "fuzzy logic," we need to develop the ability to recognize when things are "close enough" to be considered self-similar. We know how to recognize a face from a picture, and even how to discern a familial resemblance. We can recognize signatures and flavors and accents and scents, even though they are never exactly the same.

Just as the ability to learn new languages and customs decreases with age, it should follow that we lose the ability to make these sorts of recapitulated identifications as we get older. We can only imagine what the fetus remembers of his nine-month journey through evolutionary history. Most of our deepest-felt allegiances and self-similarities are forged in childhood, when we learn our relationship to the family unit, our nation, culture, and species. As children, we are wholly dependent on others for survival, so we must learn to effect change and get our needs attended to by making movements on our own level and hoping they trickle up to our elders. The better we are at recognizing our relationship to the rest of the world, the more we get what we want.

Maybe this is why the more evolutionarily advanced a species is, the longer its offspring's period of helplessness. Most species of fish and reptiles are on their own from the moment they are born. A baby deer can walk within a few hours of its birth. A human, by contrast, is relatively helpless for years. In a self-similar fashion, as we evolve culturally, the "childhood" of our kids is getting longer, too. They are socially and financially dependent on us for a longer period than we were on our own parents. Adolescents today are entertained by media intended for much younger children. We bemoan their prolonged immaturity as a sign of weakness when we could be seeing it as an indication of a higher level of social evolution. For centuries, fathers have complained that their sons grow up later and "softer" than they did. In modern times, puberty may be coming earlier, but adolescence as a developmental stage goes right through the college years. Many kids today move back home after their schooling is done and postpone marriage until well into their thirties. Although we cite economic factors for this retreat, there are social reasons as well. They want to live like kids longer in order to have more time to play. By delaying the push out of their cultural womb into self-reliance, maybe our kids are getting a longer exposure to the evolutionary blueprint of culture. They are gestating something. Germinating.

Kids may learn languages better than adults, but what if a language—like recapitulated comprehension—is so advanced that it can't be learned in five or ten years? Stay a kid long enough to learn it.

In this self-schooling, a kid must, like a developing fetus, pass through our entire cultural evolution before pushing the envelope any further. The more cultural stages to recapitulate, the longer this process will take. To become fluent in the language of chaos, our kids have enrolled in an extended graduate program lasting well into their twenties. They are teaching themselves to relate to culture and existence the way Brecht would have us relate to the stage: with meta-awareness. To do this, they need to experience and comprehend fully the self-similar quality of life today. It may prove the only alternative to confusion and despair as our world gets even more chaotic than it already is.

Rising to the occasion, as always, is "kids" television. By far the largest portion of a screenager's television diet is repeats. Classic shows in syndication, such as *I Love Lucy*, *The Dick Van Dyke Show*, *Bewitched*, *The Brady Bunch*, *Mary Tyler Moore*, *Welcome Back Kotter*, *M*A*S*H*, *Married with Children*, *Roseanne*, *The Simpsons*, and *Seinfeld* recapitulate television history in a single afternoon. Along with watching the changes in styles and scripting from each era, young viewers also glean the changing attitudes toward issues like women's roles in the workplace, race relations, sexuality, and marriage. It amounts to a fairly complete education in modern cultural history and, more important, a time-condensed recapitulation of the evolution of media.

Interestingly, the most popular shows in reruns tend to be the programs that played during the early childhoods of the teenage and twentysomething viewers. Like touchstones, the old series are frozen moments in time, giving screenagers a sense of continuity and grounding. The kids watched these shows when they were growing up and used them as lesson plans in their own social development. These classic programs taught them the basic alphabet of the language of media. Now the shows serve as wellsprings of references and associations, all framed into parody by the self-conscious, almost ritualized context of a syndicated repeat.

The joy of watching these shows, especially in tandem, is to recognize the many cultural references, while realizing how far we have come from the sensibility that spawned them in the first place. When a show from just five years ago can seem like quaint ancient history, the young audience is reassured about their ability to cope with the tremendous changes they have experienced in such a relatively short time. By coming to perceive the self-similar quality of these shows—*Murphy Brown* and *Mary Tyler Moore*, for example, both about women working in a newsroom—they have an opportunity to recognize the patterns underlying the chaos of the mediaspace and the society it represents.

Newer programming aimed directly at the screenage sensibility incorporate this recapitulated style into the format of the shows themselves. Cable TV's Comedy Central series, *Mystery Science Theater 3000* is an ongoing lesson in recognizing recapitulation, and generates comedy out of this otherwise daunting task. Conceived and launched on the fringes of mainstream media by an independent television production company in Minnesota, the program has risen to cult status and beyond among a wide variety of viewers in the pesky 18-to-34-year-old demographic. Set in the future, the show allows the audience to watch along as the sole captive inhabitant of a space station (the janitor at an invention company who was shot into space against his will) and his two robot companions are forced to view bad B-movies and low-budget science fiction sagas. Our television screen shows us the movie, with the heads of the three audience members in silhouette, Loony Tunes style, several rows ahead of us. The trio make comments and wisecracks about the movie, much as we might do if we were watching with our own friends.

But we're not. For the most part, viewers of this late-night show are isolated in their apartments, using the images on their screens as surrogate companions. In a self-similar fashion, the character trapped in the futuristic space station has fashioned his own robot friends out of spare projection parts—the ones that could have given him some control over when the movies are shown. He uses the technology at his disposal to provide himself with simulated human interaction, but has given up a certain amount of freedom to do so. So, too, do the young viewers of the show simulate a social setting with their television sets, suffering through the long, awful sci-fi movies delivered on the network's schedule, for the joy of simulated companionship. *MST 3K*, as its fans call it, both provides and recapitulates the experience of its viewers.

Do the viewers appreciate the show's efforts at recapitulation? Indeed they do. Nearly all of the robots' commentary on the bad movies they watch calls attention to the self-similar, self-reflexive, and self-conscious quality of the film-within-a-film structure of the show. Most of the films' dialogue is drowned out by the antics in the audience, and the story is lost to the endless succession of jokes and mimicry. In this otherwise disjointed and alienating world of media mirrors, recapitulation is the basis of context and comedy, lending entertaining reassurance to the young members of the observing but physically isolated media culture. The linear progression of the film's story is sacrificed to the more pressing need for a framework that recapitulates the viewing experience. In further self-similarity, the movies screened on the show often depict space travelers and other science fiction heroes who must cope with the implications of futuristic technology. Perhaps unintentional, but again self-reflexive, the man in space was himself the janitor of a high-tech invention company—the person charged with cleaning up after technological innovation.

The individual jokes and asides of the characters also comprise a new media education for the show's audience. Almost all of the humor is derived from references to other media. The "Marvel Universe" from which these characters draw their references consists of the entire pantheon of television, comic book, film, show business, and magazine history. The robots make an "Andrew Lloyd Webber" grill to burn the composer's self-derivative scores and argue about the relative merits of the Windows and Macintosh operating systems. When they observe Bela Lugosi taking off his lab coat in a campy old sci-fi feature, the robots sing, "It's a beautiful day in the laboratory" to the tune of the *Mr. Rogers' Neighborhood* theme. When a woman in a film throws down a magazine in disgust, a robot pipes "I hate Hugh Sidey!" (a correspondent for *Time* magazine). In the spirit of Brecht, and counting on the film savvy of their audience, the robots make sure to call attention to every cheesy special effect and structural flaw. As the noise of guns and guard dogs pursue escaping convicts, a robot shouts, "Sounds like the foley artists are chasing us. Move it!" Toward the end of another film, one robot comments, "Isn't this a little late in Act Three for a plot twist?"

To appreciate the humor of the show, viewers need to understand the media as a self-reflexive universe of references, any of which can be used to elucidate any other. Each joke is a demonstration of the media's self-similarity. This is not a humor of random association, but a comedy of recapitulation where images and ideas from very disparate sources are revealed as somehow relevant to one another. To belong to the *MST 3K* culture is to understand at least a majority of the literally hundreds of references per show and, more important, how they relate to one another. When this is not the object of the game, the characters instead keep their audience aware of their moment-to-moment relationship to the media, either by commenting on the technical quality of the film, or by calling attention to themselves as recapitulated bracketing devices. This commentary is further recapitulated by the hundreds of conversations occurring on the Internet at every moment in Usenet groups dedicated to *MST 3K*, where fans comment to one another about the robots' commentary, often in real-time during the broadcast!

Kids much younger than the 18-to-34 viewing demographic of *Mystery Science Theater 3000* are also getting immersed in the deeply recapitulated media of screens within screens. The Saturday morning kids' show *Reboot* takes place entirely within a video game console. The characters are programs within the computer, who must play against the invisible "user" when he inserts a game cartridge. "They say the User lives outside the net and inputs games for pleasure," explains the show's hero in the opening credits. "No one knows for sure, but I intend to find out." The characters and settings are all drawn in high-quality computer animation— they are not like computer animations, they *are* computer animations rendered on a Silicon Graphics workstation. Within the fictional world of the video game machine, the characters

235

have their own goals and conflicts that are always recapitulated in the action of the game. Whatever might be going on among the computerized characters, when a mechanized female voice interrupts to announce "warning, incoming game" a rectangular beam of light falls from the heavens, and the world changes to the setting of a game. There, the story is continued and elevated within the context of whichever game the User has chosen to play.

In one episode, the two main characters argue about whose style of life is more appropriate: Dot, the girl, plans and organizes everything, while Bob, the boy, lives on impulse. As the computer characters put it, which is better: to be "preprogrammed or random"? Fittingly, the game that they must play together against the User is a Dungeons and Dragons–style fantasy role-playing game. In a race against the User, they must move up through different levels in the game, each one requiring them to solve a puzzle in order to move on. Alternatively, they solve the game's problems using their very different life strategies, until by the end, they need to work together in order to get to the last level. As we've already seen in our exploration of the culture of chaos, neither a completely preprogrammed (deterministic) or totally random (self-determined) strategy can work on its own.

We are left to wonder what the real stakes are in these games for the characters playing. If they were to get killed within the action of a video game, would they cease to exist? For the video game–playing audience of the show, *Reboot* is an opportunity to enjoy very realistic-looking game play from the point of view of the game itself. When a kid plays a video game, the only real mystery lies within the machine's workings. For the characters in the show, that mystery comes from the real world outside the game machine, and affects their lives in very real ways. And just as the games intrude on the characters' lives from the outside, forcing them to play out their life strategies within the context of a game, the kids watching the show can see how, for them, video games offer the same instructive role-playing opportunity from a much safer vantage point.

Young people's TV is filled with less obviously recapitulated structures, plots, and themes that nonetheless contribute to their audience's confidence navigating the unfamiliar terrain of chaos. Just to relate to some of the more obscure phrases uttered by a character on *Star Trek: Voyager*—"Does anyone see anything that looks like Neelix's lungs?"—is to recapitulate, in the space of imagination, the tremendous scientific and conceptual advances made by the fictional characters in Earth's future. All four *Star Trek* series, in fact, depend heavily on the "view screen" on the ship's bridge to encounter new aliens and phenomena. It's as if we are watching television along with the crew. The cartoon characters of *The Simpsons* watch their own favorite cartoon, *Itchy and Scratchy*, and the characters in the *Batman* movies are more concerned about the way they appear in the media within the movie than they are about what's "really" going on.

But Bart Simpson and Batman are hardly traditional media characters, so we can't expect them to behave as such. They are, like an increasing number of media icons today, cross-platform media stars, who survive with great meaning independent of their original media context. Bart Simpson, as an image on a pog, still recapitulates the experience of young slacker wiseguys everywhere. Madonna on a beach towel, CD, or *Vanity Fair* cover still expresses the material girl, sexual provocateur, Marilyn Monroe, and Andy Warhol's *Marilyn* portraits. The Batman logo has become so associated with heroism and media success that Warner Bros chose to morph their own "WB" logo into Batman's at the opening of the third Batman film, *Batman Forever*; Bruce Wayne's company in the storyline, Wayne Enterprises, used a "W" logo reminiscent of Warner's.

Iconic media stars emerged in the 1950s and 1960s, as performers such as James Dean, Marilyn Monroe, Elvis Presley, and Jim Morrison achieved something akin to brand identification, independent of any single movie role or record album. Their identities remained consistent from project to project, as audiences constructed a metaplot spanning each performer's entire career. It's probably not coincidence that these first few media icons all died tragically. Unable to fully differentiate between their iconic functions and their real lives, these young stars ultimately succumbed to the traditional story arcs we were all using to understand them. Their lives needed endings. Today's pog celebrities no longer need to conform to this linear sequentiality. Be they Bart, Batman, or a real-life Madonna, they exist more as pieces of independent code that can unravel in any media context. Like a virus or a compressed computer file, they expand to recapitulate an entire battery of ideas and references.

Such iconography is mixed and matched in a variety of social milieus, from skateboard culture, the World Wide Web, comic conventions, card trading, and pog matches to Nickelodeon and MTV. Recapitulated images are so useful in a chaotic culture because they instantaneously express a set of ideas that don't need a whole linear story to be comprehended. They are moments—shapes that resonate because of their self-similarity to other moments and ideas in the culture.

Kids today self-consciously exploit this resonance in their quest to experience impending issues and ideas without the imposed structure and morality of hand-me-down linear stories and reductionist metaphors. The rave, for example, is a club experience designed to recapitulate birth as a self-consciously directed activity toward what they would call "the strange attractor at the end of time." There are no bands on stage or rock stars to cheer and emulate; this is not a story being told, but an event being recapitulated. The self-referential world of the World Wide Web also recapitulates the self-determination inherent in the chaotic experience. A screenager creates his own identity, based as much in what he points to with the links on his home page as anything he may have to say by himself. Constructing one's page is delivering oneself, as he chooses to be known, to the society at large. However naïvely conceived or psychedelically enhanced, this is the self-birth of the children of chaos.

The Whole World in Our Hands

Kids are busy ritualizing and recapitulating self-birth because we need to get some practice. Modern science and modern media both extend to us the unprecedented power of creation, and this shakes our fundamental beliefs about parenting, intelligence, and God himself.

The ability to end the life of a fetus, for example, stretches our lawmaking paradigm to its limit. Using our current moral tools we are so ill-equipped to deal with the ethical isues of this medical procedure that our entire political process can grind to a halt when a court or cabinet appointee is revealed to feel one way or the other about the matter. Many still consider the life or death of a fetus to be God's province, even though we may have developed the ability to exercise this choice ourselves. Just because we know how to do something doesn't make it "right." We have the technology to kill adults, too. People have trouble understanding why right-wingers tend to favor the death penalty while opposing abortion, and why left-wingers generally favor the right to abortion yet oppose the death penalty. It's because neither issue has anything to do with sanctioning murder. The issues are in our headlines because they test the ability (or inability) of static laws and moral templates to address adequately our power as individuals and a society over life and death. We struggle with these questions because we are looking for hard and fast rules— commandments, so to speak—that address the realities of a creature who has achieved what used to be godlike powers.

Self-determination on a personal level is almost manageable; on a societal level, it still feels incompatible with our roles as Children of the Lord.

Modern scientific research makes this even harder. We are developing new gene-splicing and biological cloning techniques faster than we can comprehend their application, much less legislate their morality. While ravers willfully "birth themselves" in recapitulated rituals, our bioengineers clone frogs and alter human genetic traits in the laboratory; we can design the qualities of our own offspring. The human invention of science, a result of our own evolutionary progress, offers us the chance to recapitulate that evolution consciously in the laboratory.

Our ability to create new types of human beings willfully, combined with our ability to destroy the entire planet with the press of a few buttons has changed our relationship to the world we were born into, whoever may have created it in the first place. Indeed, we have the whole world in our hands, and we know for whom that privilege was formerly reserved. This isn't just another Tower of Babel, where we reach up into the heavens and challenge the Lord's authority. That story depended on the idea of God as "up there" somewhere. Our metaphorical tower assumed that someone omnipotent was looking down on us mere mortals, capable of knocking us off, and dividing us up into separate peoples with incompatible languages. The enemy guardian in that myth is God himself, who stunts the human urge toward a single, self-determining, networked organism. Divide to conquer.

Well, we're back with a vengeance, and we are frightened of ourselves. But we have survived such empowering moments in history before, and we can survive this one, too. It's called a renaissance, and it's a recapitulated event—a rebirthing and interpolation of old ideas in a new context. And, like any birth, it comes because it has to.

The great Renaissance of the fifteenth and sixteenth centuries fittingly followed the paradigm-shattering development of the ability to circumnavigate the globe. When we experienced the world as planar, we quite naturally employed hierarchical models to organize our perceptions. Everything was either up or down—sky and ground, God and Earth, lord and serf. Once we fully comprehended that the world was round, our art, literature, commerce, and spiritual practices adapted to the new perception of reality. We developed perspective painting, a nod to "point of view" that represented our newly three-dimensional reality on a two-dimensional canvas. The world was no longer flat. It had depth, dimensionality, and perspective. Along with new painting techniques came the printing press and mass-produced literature, allowing people to develop their own perspectives on the Bible and other works. Caffeine from Morocco arrived, too, keeping students and intellectuals awake at night long enough to exchange and compare these new perspectives with one another.

The great dimensional shift of our own era was precipitated by our early launches into outer space. Seeing Earth from the moon gave us our first God's-eye view of the globe. The imperialist urge of Renaissance cultures gave way to a new, big-blue-marble holism. Our modern renaissance equivalent of caffeine was LSD, which led students and intellectuals to conclude that any individual perspective is itself arbitrary, and wholly dependent on the mindset of the person looking out. Our equivalent of the printing press was the computer, which, more than simply enabling us to analyze and consider data from the world around us, permitted us to express our points of view to everyone else.

Our renaissance's equivalent of perspective painting is the hologram. This optical plate allows us to record objects in three dimensions, or even over time. As the viewer changes his position, he sees a different perspective of the three-dimensional object depicted by the hologram. These different views can show the subject of picture in different positions, too. A frown can slowly change to a smile, and a bird can flap its wings. Holographic images mark another leap in our ability to appreciate the multidimensional quality of our world. For the most startling trait of a holographic plate is that if it is shattered into hundreds of pieces, each piece has the entire holographic image on it. One piece does not have one feather of the bird, and another its beak. As in a fractal, each tiny part has a faint, blurry picture of the entire image. It recapitulates the whole. When all the parts are assembled together, the whole picture comes into full resolution.

This forces us to reach a new conclusion about the multidimensional quality of our world. Each person's perspective, individually, recapitulates the whole picture, however faintly. In order to resolve our societal blur into a fully realized, multidimensional portrait, we have to let each individual perspective express itself. Everyone's point of view matters as much as anyone else's. Instead of reducing our perspectives down to simple metaphors or role models that we all must relate to, we are in the process of developing technologies that allow us all to express our personal visions simultaneously. "Stand right here when you look," the admirer of a perspective painting will tell you so that you can fully appreciate the point of view it offers. Not so in a world where the fractal represents our sense of perspective. You can look from anywhere.

Evolution itself is a journey toward dimensionality. From the blind, two-dimensional amoeba through the worm and lizard to the upright and self-conscious human, or from the simple nerve cell through the spinal cord to the ball-of-thread-like human brain, higher levels of dimensionality appear to be the goal of evolutionary development. Likewise, from the pen to moveable type to the computer, flat paintings to perspective to the hologram, or from literal meaning to metaphor to recapitulation, added levels of dimensional understanding appear to be the goal of cultural evolution. The leap to a recapitulated perspective is a pretty weird one, too. It's a matter of stepping out of the picture altogether—leaving, yet looking back. There's no real escape—just a view from a new, external point of view. An irony emerges from old biblical teaching stories like that of Lot's wife when they are reinterpreted from this recapitulated perspective. The poor woman probably got stuck when she realized it was herself burning back there in Sodom, too. You can never get out. They should have helped save their cities, and not just run away to the suburbs. But they had their orders directly from God. A recapitulated perspective was strictly forbidden in those days.

We are all fast becoming "meta," looking at ourselves from the outside while participating from the inside. Whether it's walking onto the TV screen as O. J. Simpson evades the police in the Bronco chase or splicing genes as we design the next phase of human evolution, together we are gaining the perspective of a camera shooting from over our own collective shoulder. To resolve this great portrait of six billion individual perspectives, we have begun to assemble the apparatus through which each of us will be able to experience ourselves as the creators of the world, as well as the people who have to live together in it.

Dead Metaphor: The End of Gospel As We Know It

"I'm against participatory media of all kinds," an elderly woman's voice scolded me over the phone. I was a guest on a call-in radio show, stumping for the right to Internet access when one of the long-time listeners of the heartland station got spooked. "Who will tell us how to behave?" she asked. "People will stop respecting their superiors."

"People will act how they want to," I told her.

"But people are sinful!" she responded.

"No they're not," I answered.

"Yes they are. And I'm going to dedicate my life to stopping people like you who spread this media stuff all around."

"So you're against participatory media?" I asked, trying to turn things around.

"Yes I am," she said.

"You're against people expressing their own opinions through interactive media?"

"Absolutely. It should be made illegal." She sounded almost sweet.

"Then why have you called this show, and expressed yourself over the radio?" I asked her.

There was silence on the other end of the line, then a click.

The next caller claimed to be a devout Christian.

"If people aren't evil, then why did God sacrifice his only son for our sins?" the angry man asked.

"Huh?" I was a bit confused. "Maybe people aren't so bad. Maybe we're intrinsically good. Or at least okay. There's no reason to assume we're bad."

"But Jesus died for our sins," he explained. "If human beings aren't sinful, then why did Jesus have to suffer on the cross?"

"I don't know," I answered honestly. "Maybe he didn't have to?"

Oops. The board lit up like a Christmas tree, and I spent the rest of the hour fielding death threats. They called me Satan. Strangely, I wasn't afraid. I felt almost welcomed.

Many of these folks actually *like* the idea of Satan's coming. It means that the end is near, and Judgment Day is soon to follow. Resolution to the human story in on its way. They look to the new technologies on the horizon as indications of the great fall: a world monetary system, the collapse of nation states, UPC codes that somehow can be manipulated to reveal "666," mistrust of authority, violence in the Middle East, and the implementation of a global communications network are just a few of the signs interpreted by fundamentalist radicals as indications of the apocalypse predicted in the Book of Revelations.

How could a religion be twisted to lead its believers to hope for the end of the world? By employing linear stories to quash evolutionary tales and by using static metaphors to represent our profoundly recapitulated reality.

Bible stories can prove problematic when they are applied to the modern chaotic experience. While they may not originally have been intended as closed-ended proofs of particular doctrines, they have been exploited over centuries for their potential to serve as lessons in a fixed morality. Like all stories, they have beginnings, middles, and certifiable endings. If the story of our existence has a definite ending then so, too, must we. Without any perspective, the Bible can be reduced to the pitiful: God created the world and we screwed it up. He let it collapse, but he'll grant us an exit visa if we accept his son as our personal savior. Expressed this way, it amounts to a global contract with an exclusivity clause; those who meet the terms slip out on doomsday.

The idea that we can actually do anything to make our whole world a better place contradicts the basic tenets of such apocalyptic thinking. To even try to avert the apocalypse is to go against the will of God. The object of the game is to isolate oneself from the sinners. Hole up in a bomb shelter or wilderness retreat. The world is an evil place, and we must insulate ourselves before it's too late. Those already saved hope the end is near.

This is the underlying reason why radical, doomsday-aspiring fundamentalists have so much trouble accepting an evolutionary model of biology. If things really can change, then the hard and fast rules we have been depending on are no longer absolute. The premise of this static brand of religious dogma is that we were created and are born sinful. That's the way things are. There's nothing we ourselves can do to change this. We just have to watch the preordained story unfold. But evolution provides humanity with a simple escape: change. Develop. Learn to tolerate differences. Include. If we were really to evolve toward a fully networked organization, how could we exclude anyone? Whom could we label as sinners? How would we know we were blessed if no one were damned? The apocalypse we've been counting on wouldn't even come to pass.

So, as the old lady in Colorado suspects, participatory media is the tool of the devil she fears. It brings us all together, allows us to explore new possibilities, leads us to challenge authority, and promotes the next phase of human evolution. Worst of all, it allows us to express who we really are: willful and potentially creative beings. Instead of listening to authorities, we express ourselves in an uninhibited fashion, with no moral arbiters.

But whether we like it or not, our will be done. The UPC bar-coded consumer democracy that lies ahead may allow our sinfully free will to career further out of God's hands. We say we are confused and overwhelmed by the technologies before us. Our newspapers warn us of "information overload" and "media burnout." Our firebrand preachers frighten us about angering God by messing with genetics and evolution. Our cultural guardians warn us of irreversibly repressing the values of indigenous people as our global computer networks sweep across civilization.

But none of these technologies are careening out of control. On the contrary, they are more accurately reflecting our desires and priorities every day. It is not technology we are so afraid of, but our own free will.

It's as if we were approaching a disastrous moment of total gratification. We are afraid of what we want. E-mail filters and online "intelligent agents" allow us to procure exactly the information we want from online services without browsing through junk. The articles, stock quotes, and correspondence we want are scavenged by tiny roving computer programs that then bring back the materials we are interested in reading. We can actively browse the Internet in our sleep, as these intelligent agents extend our will further in an hour than we could without them in a year.

Genetics research gives us the choice whether or not to bring a fetus with certain qualities—Down's syndrome, cerebral palsy—into this world. Many people fear the implications of this ability. They argue, shouldn't God and nature decide these things? But medical advances have all but reversed natural selection in the human species, as inherited traits that would have led to death or at least fewer offspring in the wild can now be offset with technology, thus weakening the gene pool. Natural selection is no longer allowed to improve the genetic qualities of the human species. What's so wrong, then, with bringing natural selection into the realm of conscious control? We fear what we would choose. We fear that we *can* choose.

As for technology and the free will it promotes wiping out the cultures of those who aren't white, male, and Western, we have already seen how the acceleration of networking systems leads to turbulence, which in turn increases the impact of remote influences on the overall system, and even promotes the resurgence of ideas and values long thought to be lost.

In the radical fundamentalist nightmare, fat, sinful consumers will eventually get "pizza" buttons on their television remote controls. Producers and merchants will assemble detailed profiles on consumption and spending habits, and then provide us with exactly what we want when we want it, wherever we might happen to be, thanks to bar codes on our wrists and foreheads (the signs of the devil) and tiny nanotechnology computers grafted to our brains (there are already labs growing electro-reactive fungi on microchips). Everyone, everywhere will be "tempted" to say what they think, buy what they want, get what they desire, and feel what they feel.

This acceleration of sin will precipitate the apocalypse, because we will have been liberated to express our true, dark nature.

Those who would have us subscribe to this scenario exploit the parables and gospel of Christianity and other "Metaphorical Age" religions in order to confirm their opinions. The structure of these stories, like traditional TV sitcoms, makes them particularly vulnerable to such abuse by the moralizers. You denied that beggar in the road? That was Jesus. Now you go to hell. Beginning, middle, end. Is this why someone should give to a beggar? Of course not. As most even mildly progressive Christians would agree, you give to a beggar in need because in that moment you have a spontaneous urge to help another human being to whom you relate directly. That's what a "good" person is. Sure, the threat of eternal damnation can help a misbehaving child learn self-control or even enforce charity. But is this how we want to think of ourselves—as misbehaving children with God as the disciplinarian parent?

A Little Self-Control

Young people don't think of themselves that way, and they insist on developing their own moralities. Even for crusades as conservative and remote from MTV culture as the teen celibacy movement, appeals to free will and self-determination tend to work much better than teaching stories and blanket commandments. The "True Love Waits" campaign, however Just-Say-No in its conclusion, inspired 211,163 American teenagers to sign pledge cards that read "Believing that true love waits, I make a commitment to God, myself, my family, those I date, my future mate and my future children to be sexually pure until the day I enter marriage." As a symbolic recapitulation of their mass commitment and a well-planned media virus, the cards were staked in neat rows on the lawn in front of the Washington Monument in August 1994.

By appealing to live, independent choice rather than selling prepackaged ideologies through teaching stories, the advocates of celibacy hit their target market. Whether or not we agree with their choice, and whether or not we believe they were cajoled into signing their cards, the young virgins of America have been allowed to experience their chastity as an ongoing, self-motivated commitment. It has nothing to do with external punishment or damnation.

Ironically, it's because teen sexuality results largely from an ache toward conclusion—the relief of sexual anticipation—that the advocates of celibacy were forced to adopt a strategy other than the traditional teaching story, with its tension-relieving but morally unequivocal conclusion. Teaching stories depend on consequences of actions. The character does something and later gets punished or rewarded. We are to compare our own, real-life dilemma to his, and then act the same way or differently, depending upon the outcome. We are to think, "I am like that character." But the story—the metaphor—has nothing to do with real life. It has been arbitrarily arranged.

This is why the exploitation of the stories of Jesus as absolute "gospel" is so potentially dangerous and ultimately unsatisfying in a recapitulated context. When teaching stories are forced into the service of a particular doctrine, any life in them is extinguished. As all preordained truths, they depend not on their relevance to real life but on the authority of their origin. Their meaning calcifies into dogma, and their veracity is reduced to the list of "begats" that open each gospel, proving their genuine lineage. This is hand-me-down spirituality, and requires only our compliance. These are the words of the parents. The father of the Children of the Lord.

Judeo-Christian doctrine, just like every other evolving cul-

249

tural institution, began in a literal stage of development. The laws were simply stated, like the first computer commands. Moses physically climbed up a hill and got the tablets of the Ten Commandments from God. A rule book. Not stories or metaphors, but literal directions. Don't kill, don't cheat on your wife, and don't pray to idols. The Old Testament is not meant as metaphor, but as fact. God and the angels just did this sort of stuff back then.

Jesus' story and the New Testament mark the second stage in Western religion's development, and they were developed relatively concurrently with the world's other messianic faiths. Jesus is reported as saying he came to "complete the law and the prophets." That is, he transformed one-pointed laws into events we could relate to, and similarly "completed" the predictions of earlier prophets in the actions of his own life. He turned points into lines. Society had gotten considerably more complex—allegiance was divided among family, village, religion, and emperor. Too many situations were arising that didn't fall under the simple guidelines of the literal commandments. There weren't enough rules to cover every permutation. People needed to understand something about the way these rules worked, so that they could generalize them to their particular circumstances. Thus, just as the folder and desktop of the Macintosh system arose to reduce the number of computer commands we needed to memorize, the parable was born, which could be applied to many different situations.

Jesus taught with the parable because it handily reduced complex moral conundrums into simple cause-and-effect logic. "Let he who has not sinned cast the first stone" works so well even today because we don't really throw stones at sinners. Think of how many individual rules we would need to apply the morality underlying this parable to every situation that could arise. Of course the prostitute getting stoned was someone with whom Jesus actually came in contact. It is to be considered fact, not fiction. But Jesus' entire life—whether real or invented—has served the same metaphorical purpose as the stories he told. It is the story of a parent, God, who sacrificed his son for the sins of the world. It's not that the literal stage of religion was wrong, or that the Ten Commandments were too specific. We were just too weak to apply them to our increasingly complex lives. In its primitive, metaphorical context, Jesus's death compensated for our inability to follow the literal word of God.

As with any metaphor, the Gospel requires us to make a leap of faith. *Metaphor* derives from the words meaning "to carry across." This, according to the messianic metaphor, is any savior's very reason for living and dying: to carry us across the void between life on earth and the afterlife in heaven. Jesus was reduced to and exploited as a metaphor in order to appeal to the childlike sensibility of early Christian audiences. He was used to give us a way to relate to something too difficult for us to understand directly. This is why the story of his life as depicted in the Bible so perfectly "completes" the law and the prophets of our literal Old Testament. The Passover seder becomes the Last Supper. The unleavened bread of Jewish enslavement becomes the transubstantiated, fully risen body of Christ. His story completes the law and the prophets by allowing us to draw lines between points in the Old Testament and points in his story. Religion becomes linear. It can tolerate motion, a certain amount of cultural development, and the passage of time.

When it is used as a closed-ended metaphor, the Christ story depends on identification with its hero and a tremendous leap of faith. The metaphor can carry you across to the other side, if you have *faith* in its relevance to your own story. This is the same faith that allows a child to believe his mother that the stove is hot and dangerous. At first, the mother simply says, "No." The child obeys the literal command. Eventually, the child asks, "Why?" The mother tells a story about another boy who once touched a stove and got terribly burned. The parable works. The good child unquestioningly obeys until he is old enough to determine for himself whether the stove is turned on or not. His later, adult decision to use the senses at his disposal to judge for himself the extent of the stove's threat is not a flagrant rejection of his mother's love or an attack on the genuineness with which she exercised her authority. It's just growing up.

As civilization grows up, it is becoming decidedly less linear and hierarchical. A rule like not touching the stove, which in the past was best followed blindly in all situations, loses its universally appropriate application. Just as the parabola relates an entire line to a single point, the best of parables are too often being used to relate the entire line of humanity to a uniform moral code. As our world gets more chaotic, it's getting harder to put all of humanity on a single line, however long or curvy it may be. Real-life surfaces don't conform to Euclid's smooth equations, and real-life situations don't conform to a parable's oversimplified association. And thank God they don't. When the New Testament is interpreted only as a simple parable, it forecasts the end of the world. People lose their faith and chaos reigns. The children who stay good and listen to God the parent, even though he appears to have gone away, will be rewarded when God comes home. The others will be punished forever.

But maybe we don't have to let the world degenerate into an anarchic mess as in *The Lord of the Flies*. Without their parents and authorities, the shipwrecked military school boys went mad, but they got rescued at the eleventh hour. Will we? And even if we are going to be rescued, why must we let everything fall apart before then? If Daddy's really coming home, why don't we show him how much we've grown and how well we can take care of ourselves?

Because, unlike our children of chaos, we can't take the suspense. Screenagers are willing to accept Marvel Universe creator Jack Kirby's progressive cosmology: If human beings evolve into a truly cooperative civilization, they will be rewarded with citizenship in the community of the universe. If not, poof. We have a role in our collective destiny.

As rational, pre-chaotic adults, we believe that each of us only has a role in his own personal destiny. We are not parts of a global organism. The experience of the whole is not recapitulated in the experience of each individual. Each of us is judged by God individually. Since there is no such thing as evolution, the world is doomed no matter what. Because we are born sinful, if we attempt to move beyond unquestioningly automatic faith into true, spontaneous appreciation of one another, we can only fail miserably. Without supervision, we have no choice but to rape and pillage one another. These are our natural desires. We are bad. Even the best of us are bad. The only positive role for our free will is to choose not to employ it. It will only brings things toward chaos.

The exploitation of metaphor for the instillation of fearful obeisance demands a horrible ending. A retribution. It is linear in form, hierarchical in origin, devoid of reason or experience, reductionist in its application, and absolutely singular in its conclusion. We are not allowed to grow up. The Bible quite correctly forecast this moment in history as the end. But it's not the end of the world. It's the end of the linear stories we have been using to explain our world. The story is over. The world keeps on developing.

If we cling to the metaphors we were given as children, we will surely perish. They no longer adequately describe our personal or global experience. Like a child who stays under his mother's skirt too long, we are in developmental stagnation. Like an overdue fetus, we are becoming toxic to the planet that has mothered us. It is time we gave birth to ourselves.

This is not the end of the world. It's only, as Michael Stipe,

253

of the rock band R.E.M., says, "the end of the world as we know it." And, according to the song anyway, he "feels fine." Or, as pop group Jesus Jones puts it, we are simply "watching the world wake up from history." Huh? It's the end of the linear story we have been telling ourselves. At least the MTV musicians—and their young audience—seem to understand what's going on.

Today's epidemic of apocalyptic thinking, from Ronald Reagan and Jerry Falwell to Claire Prophet and David Koresh, is a symptom of our inability to reinterpret and recontextualize the metaphors we have been using so successfully for the past few centuries. As if at the movies, we have gotten lost in the story. We are numb, and scared to wake up. Fundamentalist leaders would keep us in this state of suspension. Whether it is the Ayatollah sentencing Salman Rushdie to death for a phrase in his book or Jesse Helms lobbying for restrictions on Internet transmissions, our surrogate fathers understand that technology and communication erode the sanctity of our allegiance to absolutism. Ideas that make us think are the catalysts of chaos and the enemies of stasis. The pogs, newsgroups, and rock videos that carry them to us and our children are either the agents of our evolution—or, if the screenagers are wrong, the satanic antagonists to our original and fixed creation.

But we have to make a choice. Either abandon the templates we have been using to understand our world and restrict our behaviors, or perish in an all-consuming apocalypse and hope for a divine intervention. Whether they follow Claire Prophet prophesying the end of the world by nuclear disaster or Minister Farrakhan predicting a devastating race war, many people would rather fight, and even die, than switch paradigms at this late juncture.

But there are many of us alive on Planet Earth right now who would prefer to foster the continuation of human culture for ourselves and our children and are looking for ways to cope with the increasing complexity, discontinuity, and parentlessness of the modern experience. Our metaphorical appreciation of reality has served us like a womb, protecting us from the elements until we were ready to be born as a networked culture. It won't protect us any longer. Our chaotic kids are already working their way toward the birth canal. We owe it to them and ourselves to at least consider evolution as an alternative to death.

6

———❖———

THE FALL OF GOD AND THE RISE OF NATURE

I think there was a trade-off somewhere along the line. I think the price we paid for our golden life was an inability to fully believe in love; instead we gained an irony that scorched everything it touched. And I wonder if this irony is the price we paid for the loss of God.

But then I must remind myself we are living creatures—we have religious impulses—we *must*—and yet into what cracks do these impulses flow in a world without religion?

Douglas Coupland, *Life After God**

As frightening as it may seem, a world of chaos is a world without God—at least not as "he" is currently understood by our more evangelical and fundamentalist faiths. From the novels of Douglas Coupland and the songs of R.E.M. to the

*Douglas Coupland, *Life After God* (New York: Pocket Books, 1994), p. 273.

high-tech films of Steven Spielberg, we are immersed in a popular culture that is already reckoning with the fact that humankind must accept its role as master of its own destiny. Naturally, they all express great fear and great longing.

R.E.M.'s Michael Stipe claims he "feels fine" about "the end of the world as we know it," but in another song pines genuinely for the fact that he is "losing my religion." His lyrics in the latter song show how his new recapitulatory sensibility serves to distance him from his childhood attachment to a simple parental God. "That's me in the corner / that's me in the spotlight," he explains, as if looking down at himself. He is coping with the stress of ironic distance. He is recontextualizing, and it hurts.

Steven Spielberg and Michael Crichton's box office smash *Jurassic Park* dealt directly with man's ability to play God by creating dinosaurs out of old DNA. The scientists broke the rules of evolution by bringing back creatures from a different era. They paid dearly for the crime against nature when the monsters went out of control and tried to eat them.

Even Douglas Coupland, reluctant champion of Generation X, worries in the above passage from *Life After God* that the price of liberation from a parental god is a purgatory of irony. Ironic distance is something to be suffered—a sad side effect of breaking the spell of deference. And ironic distance kills love, which apparently can only be felt by those nostalgic, blind, or ignorant enough to still believe in it. As in Eden, knowledge breaks the trance of happiness. We may become self-conscious, but this only means we have to cover our shameful bodies with clothing and leave the Garden forever.

Our initial reactions to an apparently godless world are fear and disconnection. We become conscious of ourselves as in charge, but alone. As Coupland recognizes, this doesn't change the fact that we have religious impulses. Deep down, we know there must be a way to reconcile our disillusionment with our need to love. Our self-conscious awareness does not negate our sense that we'll only be truly fulfilled when a loving spiritual force "flows," as Douglas Coupland hopes, into the "cracks" left behind by our disillusionment.

It is a sad moment, indeed, when a child realizes his parents are not gods. It is just as sad when a civilization realizes its gods are not parents. But both facts must eventually be accepted if we are to grow into adults, and neither realization stands in the way of appreciating both parents and God for what they are.

The implicit screenage trick, if there is one, is not to abandon either parents or God just because they appear to have failed us. By using recapitulation rather than resisting its seemingly cold and dispassionate perspective, we can revivify our waning sense of spiritual order. We only have to allow our understanding of religion—as well as the Bible and Christ, if we choose that model—to evolve toward its recapitulated stage. Then it will be just as appropriate to our modern experience as its earlier stages were to younger civilizations.

We have already seen how the Judeo-Christian religion passed through literal and metaphorical stages, and how now, at the end of the metaphorical phase, its fundamentalist applications appear to be failing us as we attempt to evolve past linear and correspondingly apocalyptic thinking. Comparative mythologist Joseph Campbell argued that we are in need of a new kind of myth for the modern age. We are indeed, but maybe we don't need to dispense with the old ones altogether. We only need to experience them in modern, recapitulated terms. Just as a child only fully experiences his parents' confusion and despair when he has a child of his own, we can only come to understand the apparent failings of our gods when we experience godlike characteristics of our own.

This means looking at religion as our children of chaos would. Its value is not its authority as gospel, but its relevancy to any given moment. It doesn't work in linear time or through gravity, but as an aid in promoting an ongoing confi-

259

dence in pattern recognition and decision-making. It deconstructs the essential, repetitive shapes in the human experience so that we may apply them, as if in a fractal, on any other level in our lives.

A fantasy game–player's take on the stories in Genesis, for example, might look very different from an old-school preacher's interpretation. God blessed Abel's offering and not Cain's. For no reason. We know this feeling; things are not fair. It's less important to understand that Cain dwelled on this and turned evil and killed his brother and was damned, than to relate to his moment of despair. It has a shape, texture, and emotional reality to it. This shape and texture is so extant even in the culture of chaos that Camarilla game mythology holds that every vampire is a direct descendant of Cain himself. They are all damned, but must go on anyway. The role-playing game is based on the Bible, but gives players a chance to experience and then cope, through ritual, with a state of being that earlier biblical teachers would just tell us to avoid.

And why did Cain go mad? Because he attributed parental motivations to essentially chaotic events. God was not his father disapproving of his offering. God was Chaos, and shit happens. Two people can apply the same effort and, for no reason other than the whimsy of the dynamical system in which they exist, get very different results. One stockbroker can make a killing one day while his partner loses his shirt, even though they have used the same rule book. The losers simply need to learn from their experience and move on.

This is not necessarily satanic, for God is not the enemy. The stress and pain comes from insisting on God being a parent, rather than an embodiment of nature.

The Christ story addresses this aspect of the religious experience quite directly. If we view Christ's life as a recapitulation of the spiritual quest rather than a metaphor for God's love of his children, we come to very different conclusions about how to use the myth (or the self-similar quality of reality, if you prefer) to our advantage. Christ spent much of his life as the ultimate child of the Lord. He *knew* he was the son of God. But in order for him to move on, he needed to become a child of chaos instead. In two of the Gospels, at the moment before his death he is reported as crying out to God, "Father, why have you forsaken me?" He realizes that God is not behaving as a father to him, and experiences a moment of true disillusionment—the same moment we are experiencing as a culture today. The end of Christ's life recapitulates the end of our ability to use his story as metaphor and the necessity for us to graduate into a parentless world.

These are only examples of ways to understand the Bible in a recapitulated fashion. They are not proposed as definitive interpretations of biblical text, but rather efforts at demonstrating Christianity or any religion's ability to serve a recapitulated function rather than just a pointedly metaphorical one. We need not "lose our religion" altogether. We only need to experience it in a way that doesn't stunt our own natural evolution.

This view of religion isn't even that new. The transcendentalism of Thoreau or Emerson expressed the need to experience God as a living force—an element of nature that breathes through every object with which we come in contact. The spirit of God is recapitulated in every leaf of the forest and cloud in the sky. Only by admitting the open "cracks" in ourselves do we permit that sensibility, or love, to get inside. In this very humble manner, by accepting our frail but ultimately godlike essence do we develop from children into adults.

Whatever spirituality we accept, and the preceding paragraphs merely indicate something of my own stumble through a postatheistic quagmire, it is clear that the acceptance of our roles as co-creators of this world need not bring with it the despair of a nihilist or the egotism of a despot, but a new appreciation for the immense power of nature and our individual parts in its unfolding. The implosion of our personal space by worldwide media, as well as the nearly infinite daily reminders of our relationship to one another as members of a single, networked being give us the chance to experience a direct connection with the forces of nature without the intermediary parental devices of metaphor or authority. A new, Gaian sensibility emerges from corners as remote as the rave or the Internet, where a potentially existential and godless dilemma reveals itself as an opportunity to liberate from centuries of fear-based and paranoid social structures.

By losing God the father, the children of chaos lose the restrictive but reassuring belief in damnation, evil, Big Brother, and absolute sin. They gain a willingness to partake in their own evolution on a personal, societal, and—if the fractal holds—global level. They can cope with the fact that a chaotic reality neither guarantees them a shotgun seat next to the messiah on the highway to heaven, nor damns a thinking individual to eternal suffering in a universe allied against him.

Their new forms of media, ritual, and play instill them with the self-confidence to make personal navigational choices without fear of their own, evil inner natures. Their cross-platform, iconic cultural language gives them the ability to recognize spiritual shapes, if you will, from a variety of different sources and levels. The new mythology that Joseph Campbell foresaw may not be a new set of symbols at all, but rather a new, recapitulated appreciation of the many we already have. Even his own books comparing the myriad of spiritualities developed by different cultures around the world serve as a valuable, if academic attempt to identify the basic repetitive shapes in global mythology.

In a culture of chaos, we lose our single-minded allegiance to absolute archetypes and the definite comfort they provide. As an alternative, we must learn to identify those archetypes throughout our culture, and even in ourselves. Instead of becoming reassured in the knowledge that we have chosen the "right" path or religious institution, we seek solace in our ability to recognize the self-similar qualities essential to many different spiritualities. Today, the same kid might be tattooed with both Celtic runic and Taoist I Ching symbols, because he recognizes the validity of both and the commonality between them.

The hybridization of spiritual icons occurs across formerly distinct cultural expressions. The division between high and low culture vanishes as the sacred finds its expression in the most vernacular forums. Just as the miraculous rejuvenation of a forest after a long winter may have inspired Thoreau, we can now find reassurance in our popular culture's ability to regenerate lost icons and concepts. Although adults may be shocked and disgusted by T-shirts equating Kurt Cobain with Jesus Christ or a rave organized as a mass by an ordained Catholic priest (and both exist), these are the ways that the new, recapitulated mythology is expressed in the very natural playground of screenage culture.

Vogue, Camp, and World Peace

Despite my occasionally breathy optimism, I have my doubts, too.

If we want to find an example of what happens to a society that loses its parental gatekeepers, we need only look as far back as Bosnia or Eastern Europe. These populations did not blossom into fully realized and tolerant cultures. They have gone out of control. Racism and ethnic warfare rage. Like bugs that have been living under a rock, when that rock is pulled away they cower from the sunlight and look for something else to hide beneath. Where's the evolution?

We can't be too hard on these people. Their vigorous retreat into nationalism and xenophobia may be less a reaction to their newfound freedom than to their long-term isolation and repression. If you take a goldfish that has been kept in a tiny bowl and release him into a lake, he will swim in tiny circles for quite a while before he realizes he has more room.

A woman who has been abused by her husband may spend months with a therapist before deciding to get divorced. But when she leaves her husband, does her psyche instantly change? Usually not. Often, she will only begin to remember earlier abuses inflicted by her father or another relative. Her nightmares and distress will continue until she is able to reckon with these memories. Even after she has released her childhood abuse, she may begin to recapitulate her abuse-motif with her therapist, "transferring" the abusive role onto him in the safety of the clinical environment. Only then (assuming she has a decent therapist), after coping with these feelings and experiencing herself as the source of her own discomfort, will she be able to begin life free of the abusive cycle.

In Ukraine, just months after the demise of the Soviet Union, extremely racist SS-like groups gained tremendous popularity and power among the newly liberated people. It was as if their parents, though cruel and controlling, had suddenly abandoned them. Unprepared for adulthood, the child-like populations immediately resorted to their last memories of a paternal regime, however restrictive its rules. They were willing to commit themselves to another repressive regime, as long as someone agreed to parent them.

In 1995, on the day that the fascist Italian Social Movement formed by Mussolini's last followers finally disbanded, one of its former deputies explained to the *New York Times*, "It was our family. Today we are abandoning our father's house and moving into our own. Tonight, I thought of my father and all these emotions came rushing out."* Luckily for men like this deputy, a new Fascist party called the

*Celestine Bohlen, "Neo-Fascists Remodel Their Party in Italy," *New York Times*, 30 January 1995.

National Alliance has arisen to fulfill the urge for the certainty of paternal politics.

Just because our first reaction to freedom is to find new ideological parents to replace the old ones, we must not conclude that we have an intrinsic need for restrictive governance. Nor should we condemn an individual to an infinite succession of abusive relationships just because she has a bit of trouble relating to people in a different way. She simply needs therapy of one kind or another. There, the woman recapitulates her abusive cycle in the ritual we call psychoanalysis as a means of liberating herself from the pattern itself. Similarly, societies retreating into cults and militias as their former institutional dominators decline should not be dismissed as doomed to repeat the same mistakes forever. They, too, are in need of social therapy—some way of recapitulating their cyclic behaviors under safe conditions that allow for observation, insight, and adjustment. This therapy is called play.

The media, ritual, and fantasy of the children of chaos are the activities arising quite naturally to reassure us of our ability to survive our collective destiny. Our long-standing ideologies, cultural icons, and systems of morality are all being reevaluated in their new chaotic context. The ones that work well get spread further and better than ever before, while the ones that don't work just fade into oblivion. It's as if our media culture were a giant machine—a framing machine that brackets and rebrackets every idea, issue, and utterance. By the time the original concept—be it Nazism, Nike, or the nation-state—is cycled through the machine it has either been bracketed into laughable obscurity or disseminated globally.

Things are bracketed, or made "meta," in a number of ways. Gay values are often bracketed as "camp." Transvestites couch their culturally dangerous ideas in self-consciously posed, or vogued, versions of favorite female icons. *Beavis and Butt-head* and *Mystery Science Theater 3000* bracket inane media in screens within screens. The threat of neo-Naziism is repetitively framed in the hall of mirrors known as the American talk show, where the Nazis' empty platitudes are eventually reduced to the eloquence of "yo momma." Meanwhile culturally relevant comic book heros, the Gaia hypothesis, and the Internet serve as propagators of valuable cultural catalysts. Despite well-funded disinformation campaigns by cigarette companies, the University of California used the Internet to post documentation of the tobacco industry's concealment of smoking's harmful effects. The empowering messages of recording artists are iterated millions of times on CDs and videos that can penetrate the best-fortified cultural boundaries, inspiring revolution in South Africa and revolutionary thought in suburban high schools.

A chaotic culture works like an iterative equation. It is a turbulent, dynamical system. A fractal. Anything generated by it is plugged back in, again and again, until it either changes the whole system or dissipates into nothing. In a chaotic cultural system, ideas generating tolerance, better survival, or higher levels of organization will continue to iterate. Those that don't find enough resonance simply fade into oblivion, like random but ineffective mutations in a species of bacteria.

As our culture becomes more turbulent, it becomes more stable and effective, not less. Like a living thing, the chaotic culture is self-regulating. Ideas that improve the overall stability and development of the system will replicate, while ideas that don't will naturally iterate into obscurity. Others— like animé or Madonna—will work and multiply for time, until their use has been realized. In the worst cases, like UFO paranoia or even Naziism, what doesn't kill us before it is completely iterated will, at the very least, teach us something.

Endgame

Our chaotic world and the networks holding it together function as an imitation or, better, recapitulation of nature. Our fear of moving into a culture of chaos is less a fear of technology and confusion than a fear of nature itself.

New Agers would probably prefer I say "herself." They like to think of nature as a "she." It is comforting to imagine that as our paternal bonds dissolve, warm, maternal ones will rise to take their place. But the planet that spawned us is not our mother any more than God is our father. It is indifferent to our plight and will get along quite well without us. Environmentalists argue that we must quickly act in order to save the earth. Hardly. No matter what we do, something—weeds, roaches, fungi—will live on. If we need to act now, it is in order to keep the planet a viable host for *ourselves*. If the damage we inflict on the planet outweighs the stability our culture can provide, we will surely be neutralized by one of nature's many other regulatory systems. One good plague or ozone hole could pretty easily take care of us all.

This would not be the conscious retribution of an earth spirit "mother." It would be something we did to ourselves. *We* are the conscious beings here. It would be a shame, though, because we are probably the most evolved and self-conscious creatures on the planet to date. Still, the entire span of our existence accounts for a tiny fraction of the time life has been here. We are merely an attempt—a stab—at networking this whole thing together. If we fail, nature will simply try again—not consciously, but naturally—to reach a higher level of organization.

To accept our chaotic culture we must accept that nature has no prejudice, other than to promote stability and higher levels of organization. Life is simply a battle against entropy. If an organism or culture generates more entropy than organization, it is working against life, and its inefficiency will lead to its demise. Many believe we have brought our own civilization to such a precipice, but in truth the many inventions and institutions we created to resist the forces of nature are finally working in life's favor. Our media promotes free communication, our economy promotes choice, and our religions are turning to participation over programming.

To give in to the force of nature is no easy task, especially when we feel we've gotten this far by resisting it. We sense that the roles we have been playing and institutions we have been counting on are eroding, and we aren't sure that we would be "entitled" to such comforts in a more naturally resolved schema. The great "revolt of the elites" so many intellectuals write about today amounts to a series of justifications for maintaining our artificially stalemated culture in the face of its certain modification. For as an ocean slowly erodes the shape of continents, our newly turbulent culture will erode the shapes of our inefficient social structures. Our manmade inventions and institutions—be they media or military—have come full circle, and generate so much turbulence that they serve to restore nature rather than circumvent it.

We have been trained (or trained ourselves) to associate this passage into nature as the end of the world. Apocalypse. And, given the events described on our evening newscasts, it has become easy to believe it. But as many philosophers in similar periods of rebirth have explained, we may be misreading the symptoms of renaissance as those of our civilization's doom. Idealists from ancient Greece's Plato to the nineties' psychedelics advocate Terence McKenna have compared these societal shifts to labor pains—contractions before the release.

If you were to come upon a woman giving birth and you knew nothing about the way babies are born, you might assume something was terribly wrong. The commotion would include heaving, moaning, convulsing, sweating, and bleeding. Her predicament would appear pretty ghastly—even hopeless. Surely this woman is about to die.

But as we know, however painful the process of child delivery, it is usually not fatal. In fact it brings about new life. It is perfectly natural, and the more one relaxes the better it generally goes.

Such a process is occurring in our world right now. Almost every conventional cultural indicator registers disaster, but if we adopt a more evolutionary perspective toward our plight we may be surprised by just how reassuring these same symptoms appear. Everything is proceeding on schedule.

Like a fetus embarked on the journey toward birth, we must give up the comfort of the womb for the life that lies ahead. If, on the other hand, we choose to resist, we will become as toxic to our own environment as an overdue fetus does to its mother. Maybe we've even been procrastinating a bit already.

Our children, ironically, have already made their move. They are leading us in our evolution past linear thinking, duality, mechanism, hierarchy, metaphor, and God himself toward a dynamic, holistic, animistic, weightless, and recapitulated culture. Chaos is their natural environment.

By following our screenagers' example rather than panicking at their embrace of turbulence, we may just stand a chance of adapting to the culture toward which we are inevitably migrating. The thing we are about to become is already with us. Just look—really look—at your children for tangible proof, beyond the shadow of a doubt, that everything is going to be all right.

And if you still can't reconcile yourself to life in the real world, it may not even matter. Lord only knows what a fetus thinks on its way down the birth canal.

SELECTED READINGS
AND RESOURCES

Books

Briggs, John, and F. David Peat. *Turbulent Mirror*. New York: HarperPerennial, 1990.

Coupland, Douglas. *Life After God*. New York: Pocket Books, 1994.

Frauenfelder, Sinclair, Kreth, and Branwyn, eds. *The Happy Mutant Handbook*. New York: Riverhead, 1995.

Jacobs, Jane. *Systems of Survival*. New York: Vintage, 1992.

Leary, Timothy. *Chaos and Cyberculture*. Berkeley: Ronin Publishing, 1994.

Lovelock, J. E. *Gaia: A New Look at Life on Earth*. New York: Oxford University Press, 1979.

McCloud, Scott. *Understanding Comics*. New York: Harper-Perennial, 1994.

McKenna, Terence. *The Archaic Revival*. San Francisco: Harper-SanFrancisco, 1991.

McLuhan, Marshall. *Understanding Media*. New York: McGraw-Hill, 1964.

Rein-Hagen, Mark. *Vampire: The Masquerade.* Stone Mountain, Ga.: White Wolf, 1992.

Re/Search. A trade paperback book series on fringe culture.

Wice, Nathaniel, and Steven Daly. *alt.culture.* New York: Harper-Collins, 1996.

Organizations

THE CENTER FOR MEDIA LITERACY
1962 SOUTH SHENANDOAH STREET
LOS ANGELES, CA 90034

Phone:	310/559-2944
	800/226-9494
Fax:	310/559-9396
e-mail:	jdover@earthlink.net

STRATEGIES FOR MEDIA LITERACY, INC.
P.O. BOX 460910
SAN FRANCISCO, CA 94146-0910

Phone:	415/621-2911
e-mail:	sml@fwl.edu

Periodicals

Adbusters: Media analysis.
Animag: Japanese animation, in English.
Big Brother: Skateboard culture magazine.
Blunt: Snowboarding.
Grand Royal: Popular 'zine by the band The Beastie Boys.
Mecha Press: Gundam and animé shows, models, and toys.
Next Generation: Video games and innovations.

Propaganda: Goth culture and music.

Slap: Skateboarding and skate culture.

Spin: Music and youth culture.

Surfer

Urb: Rave, hip-hop, club life, spirituality, and underground youth culture.

Internet Sites

CHILDREN'S ISSUES AND RESOURCES FOR PARENTS

Children Now: Kid Advocates and other Web resources for parents. http://www.dnai.com/~children/

The EdWeb Home Room Information on Technology and School Reform, by the Corporation for Public Broadcasting. http://k12.cnidr.org:90/resource.cntnts.html

Global Child Health News and Review: Widely endorsed source of information on world kids' culture and issues. http://edie.cprost.sfu.ca/gcnet/gchnr.html

CULTURE

Hyperreal: Alternative culture and music; mostly rave and psychedelic. http://hyperreal.com/

The Schwa Corporation: A mock-paranoid art project. http://www.theschwacorporation.com/

The Church of the SubGenius: Parody cult religion. http://sunsite.unc.edu/subgenius/

The Dominion: Sci-Fi Channel's science fiction, horror, and fantasy resources and references. http://www.scifi.com/

Tumyeto Ghetto: Information and resources for youth culture enthusiasts, including skateboarders, snowboarders, and extreme sports fans. http://www.tumyeto.com

Spiv: Turner Entertainment's youth culture site. http://www.spiv.com

POLITICS

Mediafilter: Good "GenX" style politics and social action. http://MediaFilter.org/MFF/mfhome

Usenet groups: alt.society.genx *and* misc.activism.progressive.

INDEX